T0261965

Oxidative Stress and Related Diseases

Volume II

Oxidative Stress and Related Diseases
Volume II

Edited by **Nick Gilmour**

New York

Published by Callisto Reference,
106 Park Avenue, Suite 200,
New York, NY 10016, USA
www.callistoreference.com

Oxidative Stress and Related Diseases
Volume II
Edited by Nick Gilmour

© 2015 Callisto Reference

International Standard Book Number: 978-1-63239-504-7 (Hardback)

This book contains information obtained from authentic and highly regarded sources. Copyright for all individual chapters remain with the respective authors as indicated. A wide variety of references are listed. Permission and sources are indicated; for detailed attributions, please refer to the permissions page. Reasonable efforts have been made to publish reliable data and information, but the authors, editors and publisher cannot assume any responsibility for the validity of all materials or the consequences of their use.

The publisher's policy is to use permanent paper from mills that operate a sustainable forestry policy. Furthermore, the publisher ensures that the text paper and cover boards used have met acceptable environmental accreditation standards.

Trademark Notice: Registered trademark of products or corporate names are used only for explanation and identification without intent to infringe.

Printed in the United States of America.

Contents

Preface

Over the recent decade, advancements and applications have progressed exponentially. This has led to the increased interest in this field and projects are being conducted to enhance knowledge. The main objective of this book is to present some of the critical challenges and provide insights into possible solutions. This book will answer the varied questions that arise in the field and also provide an increased scope for furthering studies.

This book is a collaborative effort for progressing towards the prevention and treatment of chronic degenerative diseases. It responds to the urge to find, in a sole document, the impact of oxidative stress at distinct levels, as well as treatment with antioxidants to react and decrease the damage. The book includes two sections namely, aging and therapy- a role for antioxidants. Researchers from across the world have contributed in it. The book is intended for health professionals and those interested in combating stress.

I hope that this book, with its visionary approach, will be a valuable addition and will promote interest among readers. Each of the authors has provided their extraordinary competence in their specific fields by providing different perspectives as they come from diverse nations and regions. I thank them for their contributions.

Editor

Aging

Aging, Oxidative Stress and Antioxidants

B. Poljsak and I. Milisav

Additional information is available at the end of the chapter

1. Introduction

Aging is an extremely complex and multifactorial process that proceeds to the gradual deterioration in functions. It usually manifest after maturity and leads to disability and death. Traditionally researchers focused primarily on understanding how physiological functions decline with the increasing age; almost no research was dedicated to investigation of causes or methods of aging intervention. If scientists would discover a drug for healing all major chronic degenerative diseases, the average lifetime would be increased for just 12 years. People would still die from complications connected with the aging process.

Defects formed in human body as a consequence of the aging process start to arise very early in life, probably *in utero*. In the early years, both the fraction of affected cells and the average burden of damage per affected cell are low [1]. The signs of aging start to appear after maturity, when optimal health, strength and appearance are at the peak. After puberty, all physiological functions gradually start to decline (e.g. the maximum lung, heart and kidney capacities are decreased, the secretion of sexual hormones is lowered, arthritic changes, skin wrinkling, etc). The precise biological and cellular mechanisms responsible for the aging are not known, but according to Fontana and Klein [2], "they are likely to involve a constellation of complex and interrelated factors, including [1] oxidative stress–induced protein and DNA damage in conjunction with insufficient DNA damage repair, as well as genetic instability of mitochondrial and nuclear genomes; [2] noninfectious chronic inflammation caused by increased adipokine and cytokine production; [3] alterations in fatty acid metabolism, including excessive free fatty acid release into plasma with subsequent tissue insulin resistance; [4] accumulation of end products of metabolism, such as advanced glycation end products, amyloid, and proteins that interfere with normal cell function; [5] sympathetic nerve system and angiotensin system activation as well as alterations in neuroendocrine systems; and [6] loss of post-mitotic cells, resulting in a decreased number of neurons and muscle cells as well as deterioration in structure and function of cells in all tissues and organs".

In recent years, oxidative stress has been implicated in a wide variety of degenerative processes, diseases and syndromes, including the following: mutagenesis, cell transformation and
cancer; heart attacks, strokes, atherosclerosis, and ischemia/reperfusion injury; chronic inflammatory diseases, like rheumatoid arthritis, lupus erythematosus and psoriatic arthritis; acute
inflammatory problems; photooxidative stresses to the eye, e.g. cataract; neurological disorders, such as certain forms of familial amyotrophic lateral sclerosis, certain glutathione per-
oxidase-linked adolescent seizures, Parkinson's and Alzheimer's diseases; and other age-
related disorders, perhaps even including factors underlying the aging process itself [3].

2. Aging theories

Scientists estimated that the allelic variation or mutations in up to 7,000 relevant genes
might modulate their expression patterns and/or induce senescence in an aging person, even
in the absence of aging specific genes [4, 5]. As these are complex processes they may result
from different mechanisms and causes. Consequently, there are many theories trying to explain the aging process, each from its own perspective, and none of the theories can explain
all details of aging. The aging theories are not mutually exclusive, especially, when oxidative stress is considered [6].

Mild oxidative stress is the result of normal metabolism; the resulting biomolecular damage
cannot be totally repaired or removed by cellular degradation systems, like lysosomes, proteasomes, and cytosolic and mitochondrial proteases. About 1% to 4% of the mitochondrial-
ly metabolized oxygen is converted to the superoxide ion that can be converted
subsequently to hydrogen peroxide, hydroxyl radical and eventually other reactive species,
including other peroxides and singlet oxygen that can in turn, generate free radicals capable
of damaging structural proteins and DNA [7, 8, 9, 10, 11]. Since extensive research on the
relation between polymorphisms likely to accelerate/decelerate the common mechanisms of
aging and resistance to the oxidative stress has been neglected in almost all scientific studies, the data do not allow us to conclude that the oxidative theory supports the theory of
programmed aging so far [7]. However, the most recent studies support the idea that oxidative stress is a significant marker of senescence in different species. Resistance to oxidative
stress is a common trait of long-lived genetic variations in mammals and lower organisms
[5, 12]. Theories on aging process can be divided into programmed and stochastic.

2.1. Free radical theory, oxidative stress theory and mitochondrial theory of aging

Denham Harman was first to propose the free radical theory of aging in the 1950s, and extended the idea to implicate mitochondrial production of reactive oxygen species in 1970s,
[13]. According to this theory, enhanced and unopposed metabolism-driven oxidative stress
has a major role in diverse chronic age-related diseases [13, 14, 7]. Organisms age because of
accumulation of free radical damage in the cells. It was subsequently discovered that reactive oxygen species (ROS) generally contribute to the accumulation of oxidative damage to
cellular constituents, eventhough some of them are not free radicals, as they do not have an

unpaired electron in their outer shells [15, 16]. Consistently, aged mammals contain high quantities of oxidized lipids and proteins as well as damaged/mutated DNA, particularly in the mitochondrial genome [13, 14]. In support of a mitochondrial theory of aging, the mitochondrial DNA damage increases with aging [17, 18]. Thus, a modern version of this tenet is the "oxidative stress theory" of aging, which holds that increases in ROS accompany aging and lead to functional alterations, pathological conditions, and even death [19].

The oxygen consumption, production of ATP by mitochondria and free-radical production are linked processes [20, 21]. Harman first proposed that normal aging results from random deleterious damage to tissues by free radicals [14] and subsequently focused on mitochondria as generators of free radicals [13]. Halliwell and Gutteridge later suggested to rename this free radical theory of aging as the "oxidative damage theory of aging" [22], since aging and diseases are caused not only by free radicals, but also by other reactive oxygen and nitrogen species.

Increases in mitochondrial energy production at the cellular level might have beneficial and/or deleterious effects [23]. Increased regeneration of reducing agents (NADH, NADPH and $FADH_2$) and ATP can improve the recycling of antioxidants and assist the antioxidant defence system. On the other hand, enhanced mitochondrial activity may increase the production of superoxide, thereby aggravating the oxidative stress and further burdening the antioxidant defence system. The mitochondria are the major source of toxic oxidants, which have the potential of reacting with and destroying cell constituents and which accumulate with age. The result of this destructive activity is lowererd energy production and a body that more readily displays signs of age (e.g., wrinkled skin, production of lower energy levels). There is now a considerable evidence that mitochondria are altered in the tissues of aging individuals and that the damage to mitochondrial DNA (mtDNA) increases 1,000-fold with age [24].

The mutation rate of mitochondrial DNA is ten-times higher than that of nuclear DNA. Mitochondrial DNA (mtDNA) is a naked, mostly double-stranded, circular, and is continuously exposed to ROS. It is replicated much faster than nuclear DNA with less proofreading and efficient DNA repair mechanisms [25]. Thus, mtDNA is more vulnerable to attack by ROS. Damaged mitochondria can cause the energy crisis in the cell, leading to senescence and aging of tissue. Accumulation of damage decreases the cell's ability to generate ATP, so that cells, tissues, and individuals function less well. The gradual loss of energy experienced with age is paralleled by a decrease in a number of mitochondria per cell, as well as energy-producing efficiency of remaining mitochondria.

A major effect of mitochondrial dysfunction is an unappropriately high generation of ROS and proton leakage, resulting in lowering of ATP production in relation to electron input from metabolism. Leaked ROS and protons cause damage to a wide range of macromolecules, including enzymes, nucleic acids and membrane lipids within and beyond mitochondria, and thus are consistent with the inflammation theory of aging as being proximal events triggering the production of pro-inflammatory cytokines. The age-related increases in the levels of both oxidative damage and mutational load of mtDNA predicted by the mitochondrial theory of aging have been described in multiple species and organ systems [26]. However, whether this damage affects mitochondrial function or significantly modulates the physiology of aging has remained controversial [27, 28]. As already mentioned, free radicals can damage the mitochondrial inner membrane, creating a positive feedback-loop for increased free-radical creation. Induction of ROS generates mtDNA mutations, in turn leading to a defective respiratory chain. Defective respiratory chain generates even more ROS and generates a vicious cycle. The result is even more damage.

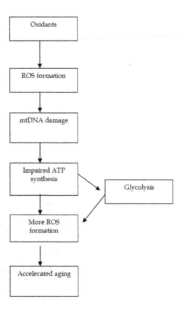

Figure 1. Oxidative stress from endogenous or exogenous sources can trigger the chain reaction, which leads to accelerated aging process of cells and organisms.

On the other hand, the "vicious cycle" theory, which states that free radical damage to mitochondrial DNA leads to mitochondria that produce more superoxide, has been questioned by some scientists since the most damaged mitochondria are degraded by autophagy, whereas the less defective mitochondria (which produce less ATP as well as less superoxide) remain to reproduce themselves [29]. But the efficiency of autophagy to consume malfunctioning mitochondria also declines with age, resulting in more mitochondria producing higher levels of superoxide [30]. Mitochondria of older organisms are fewer in number, larger in size and less efficient (produce less energy and more superoxide).

Free radicals could also be involved in signalling responses, which subsequently stimulate pathways related to cell senescence and death, and in pro-inflammatory gene expression. This inflammatory cascade is more active during aging and has been linked with age-associated pathologies, like cancer, cardiovascular diseases, arthritis, and neurodegenerative diseases [31].

2.2. Other theories of aging

Apart from the free radical theory, the aging is explained by many other theories:

The Telomere shortening hypothesis (also described as "replicative senescence," the "Hay-flick phenomenon" or Hayflick limit) is based on the fact that telomeres shorten with each successive cell division. Shortened telomeres activate a mechanism that prevents cell division [32]. The telomere shortening hypothesis cannot explain the aging of the non-dividing cells, e.g. neurons and muscle cells, thus cannot explain the aging process in all the cells of an organism.

The Reproductive-cell cycle theory states that aging is regulated by reproductive hormones, which act in an antagonistic pleiotropic manner through cell cycle signaling. This promotes growth and development early in life in order to achieve reproduction, however later in life, in a futile attempt to maintain reproduction, become dysregulated and drive senescence [32].

The Wear and tear theory of aging is based on the idea that changes associated with aging result from damage by chance that accumulates over time [32]. The wear-and-tear theories describe aging as an accumulation of damage and garbage that eventually overwhelms our ability to function. Similar are Error accumulation and Accumulative waste theories; Error accumulation theory explains aging as the results from chance events that escape proofreading mechanisms of genetic code [32], according to Accumulative waste theory the aging results from build-up of cell waste products in time because of defective repair-removal processes. Terman, [33] believes that the process of aging derives from imperfect clearance of oxidatively damaged, relatively indigestible material, the accumulation of which further hinders cellular catabolic and anabolic functions (e.g. accumulation of lipofuscin in lysosomes). The programmed theories (e.g. aging clock theory) propose a time-switch in our bodies that controls not only our process of development but also triggers our self-destruction. The shortening of telomeres would provide such a clock in rapidly dividing cells. The Autoimmune theory of aging is based on the idea that aging results from an increase in antibodies that attack the body's tissues [32].

Mitohormesis theory of aging is based on the "hormesis effects". It describes beneficial actions resulting from the response of an organism to a low-intensity stressor. It has been known since the 1930s that restricting calories while maintaining adequate amounts of other nutrients can extend the lifespan in laboratory animals. Michael Ristow's group has provided evidence for the theory that this effect is due to increased formation of free radicals within the mitochondria causing a secondary induction of increased antioxidant defense capacity [34]. Finkel et al., [35] stated that the best strategy to enhance endogenous antioxidant levels may actually be oxidative stress itself, based on the classical physiological concept of hormesis (for detailed information on hormesis see paragraph Adaptive responses and hormesis).

Additionally, the Disposable soma theory was proposed [36, 37], which postulated a special class of gene mutations with the following antagonistic pleiotropic effects: these hypothetical mutations save energy for reproduction (positive effect) by partially disabling molecular proofreading and other accuracy promoting devices in somatic cells (negative effect). The

Evolutionary theory of aging is based on life history theory and is constituted of a set of ideas that themselves require further elaboration and validation [38].

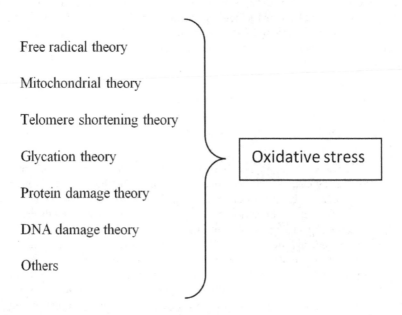

Figure 2. Oxidative stress as the common denominator of majority of aging theories.

Evidence implies that an important theme linking several different kinds of cellular damage is the consequence of exposure to reactive oxygen species [5, 39].

Many of the theories overlap, e.g., ROS can cause DNA damage (free radical theory) and also accelerate telomere shortening (telomere theory), since telomere shortening is accelerated by oxidative stress in vascular endothelial cells [40, 41]. None of the theories explain the aging process, as it may be too complex to be covered by only one theory. Perhaps there is no single mechanism responsible for aging in all living organisms [42]. The definitive mechanisms of aging across species remain equivocal. Diminished capacity for protein synthesis and DNA repair, decline in immune functions, loss of muscle mass and strength, a decrease in bone mineral density as well as a decrease in enzymatic and non-enzymatic antioxidative protections are well established. In essence, aging is progressive accumulation through life of many random molecular defects that build up within the cells and tissues. For this reason, only one "magic bullet" will never be able to prevent or reverse the complex and multicausal process of aging.

3. The Role of Oxidative Stress on the General Aging Process

In order to understand strategies to reduce oxidative stress and aging, it is first important to briefly explain reasons for oxidative stress formation. Oxidative damage is a result of the intrinsic and extrinsic ROS formation factors. The most important endogenous sources of oxidants are mitochondrial electron transport chain and nitric oxide synthase reaction, and the non-mitochondrial soruces: Fenton reaction, reactions involving cytochromes P450 in microsomes, peroxisomal beta - oxidation and respiratory burst of phagocytic cells [6]. Free radical reactions have been implicated also as the consequence of exposure to many environmental pollutants, e.g. cigarette smoke, alcohol, ionizing and UV radiations, pesticides, ozone, etc. Oxidative stress is the direct consequence of an increased generation of free radicals and/or reduced physiological activity of antioxidant defenses against free radicals. The degree of oxidative stress is proportional to the concentration of free radicals, which depends on their formation and quenching.

Causes of increased free-radical production include [43]:

Endogenous

- elevation in O_2 concentration
- increased mitochondrial leakage
- inflammation
- increased respiration
- others

Exogenous

- environment (pollution, pesticides, radiation, etc.)
- smoking
- poor nutrition
- disorders and chronic diseases
- chronic inflammation
- lifestyle
- strenuous excercise
- psychological and emotional stress
- others

Causes of decreased antioxidant defense include:

- reduced activity of endogenous antioxidative enzymes
- reduced biokinetics of antioxidant metabolism
- reduced intake of antioxidants
- reduced bioabsorption of antioxidants
- others

Oxidative stress is caused mainly by:

- mutation or reduced activity of enzymes (catalase, SOD, glutathione peroxidase)
- decreased intake of exogenous antioxidants from food
- increased metal ion intake (e.g., Fe, Cu, Cr)
- easiliy peroxidized amino acids (e.g., lysine)
- increased triplet oxigen ($^3O_2$) concentration
- increased physical activity of an untrained individual
- ROS from ionizing radiation, air pollution, smoking
- chronic inflammation

Excessive generation of free radicals may overwhelm natural cellular antioxidant defenses, leading to oxidation and further functional impairment. There is an oxidative damage potential, as there is a constant free radical formation in small amounts, which escape the cell defense.

The reduction of oxidative stress can be achieved on three levels [44]: i) by lowering exposure to environmental pollutants ii) by increasing the levels of endogenous and exogenous antioxidants in order to scavenge ROS before they can cause any damage; or iii) lowering the generation of oxidative stress by stabilizing mitochondrial energy production and efficiency - reducing the amount of ROS formed per amount of O_2 consumed.

4. Defenses against ROS and strategies to reduce oxidative stress

Generation of ROS and the activity of antioxidant defenses are balanced *in vivo*. In fact, the balance may be slightly tipped in favor of ROS so that there is continuous low-level oxidative damage in the human body.

Besides the endogenous and exogenous antioxidative protection, the second category of defence are repair processes, which remove the damaged biomolecules before they accumulate to cause altered cell metabolism or viability [45].

4.1. Primary Antioxidant Defenses

Superoxide Dismutase (SOD)

SODs are a group of metalloenzymes, which catalyze the conversion of superoxide anion to hydrogen peroxide and dioxygen [46]. This reaction is a source of cellular hydrogen peroxide.

$$2O_2{}^{-} + 2H^{+} \rightarrow H_2O_2 + O_2 \tag{1}$$

Catalase

Hydrogen peroxide formed by SOD, from other metabolic reactions or from the non-enzymatic reaction of the hydroperoxyl radical, is scavenged by a ubiquitous heme protein catalase. It catalyzes the dismutation of hydrogen peroxide into water and molecular oxygen [47].

$$2\,H_2O_2 \rightarrow O_2 + 2H_2O \tag{2}$$

One antioxidative role of catalases is to lower the risk of hydroxyl radical formation from H_2O_2 via Fenton reaction catalyzed by chromium or ferrous ions.

Glutathione Peroxidase (GPx)

All glutathione peroxidases may catalyze the reduction of H_2O_2 using glutathione (GSH) as a substrate. They can also reduce other peroxides (e.g., lipid peroxides in cell membranes) to alcohols.

$$ROOH + 2\,GSH \rightarrow ROH + GSSG + H_2O \tag{3}$$

GPx is responsible for detoxification of low H_2O_2 amounts, while in higher H_2O_2 amounts, catalase takes the leading part in cellular detoxification [15].

Glutathione-Related Systems

In addition to enzymatic defenses described above, there is an intracellular non-enzymatic defense system to protect cellular constituents against ROS and for maintaining the redox state. Glutathione (GSH) is the most abundant intracellular thiol-based antioxidant, present in millimolar concentrations in all aerobic cells, eukaryotic and prokaryotic [48]. It is a sulfhydryl buffer, detoxifies compounds through conjugation reactions catalyzed by glutathione S-transferases, directly, as in the case with peroxide in the GPx-catalyzed reaction [47] or with Cr(VI) [49]. GSH is capable of reacting with Cr(VI) to yield Cr(V), Cr(IV), GSH thiyl radicals and Cr(III)-GSH complexes [50, 51]. The ratios of reduced-to-oxidized glutathione (GSH/GSSG) in normal cells are high (> 10 : 1), as the enzyme, glutathione reductase, help to reduce oxidized glutathione in the following reaction:

$$GSSG + NADPH + H^+ \rightarrow 2GSH + NADP^+ \qquad (4)$$

The NADPH required is from several reactions, the best known from the oxidative phase of pentose phosphate pathway [15]. Both, glutathione reductase and glucose-6-phosphate dehydrogenase are involved in the glutathione recycling system [52].

4.2. Secondary Antioxidant Defenses

Although efficient, the antioxidant enzymes and compounds do not prevent the oxidative damage completely. A series of damage removal and repair enzymes deal with this damage. Many of these essential maintenance and repair systems become deficient in senescent cells, thus a high amount of biological "garbage" is accumulated (e.g., intralysosomal accumulation of lipofuscin) [53, 54]. Age-related oxidative changes are most common in non-proliferating cells, like the neurons and cardiac myocites, as there is no "dilution effect" of damaged structures through cell division [33]. The ability to repair DNA correlates with species-specific lifespan, and is necessary, but not sufficient for longevity [55]. There is an age-related decline in proteasome activity and proteasome content in different tissues (e.g. rat liver, human epidermis); this leads to accumulation of oxidatively modified proteins [56]. Proteasomes are a part of the protein-removal system in eukaryotic cells. Proteasome activity and function may be decreased upon replicative senescence. On the other hand, proteasome activation was shown to enhance the survival during oxidative stress, lifespan extension and maintenance of the juvenile morphology longer in specific cells, e.g. human primary fibroblasts [57]. The total amount of oxidatively modified proteins of an 80-year-old man may be up to 50% [58]. Besides, elevated levels of oxidized proteins, oxidized lipids, advanced DNA oxidation and glycoxidation end products are found in aged organisms [7, 59, 60]. Torres and Perez [61] have shown that proteasome inhibition is a mediator of oxidative stress and ROS production and is affecting mitochondrial function. These authors propose that a progressive decrease in proteasome function during aging can promote mitochondrial damage and ROS accumulation. It is likely that changes in proteasome dynamics could generate a prooxidative conditions that could cause tissue injury during aging, *in vivo* [61].

Numerous studies have reported age-related increases in somatic mutation and other forms of DNA damage, indicating that the capacity for DNA repair is an important determinant of the rate of aging at the cellular and molecular levels [62, 63]. An important player in the immediate cellular response to ROS-induced DNA damage is the enzyme poly(ADP-ribose) polymerase-1 (PARP-1). It recognizes DNA lesions and flags them for repair. Grube and Burkle [64] discovered a strong positive correlation of PARP activity with the lifespan of species: cells from long-lived species had higher levels of PARP activity than cells from short-lived species.

The DNA-repair enzymes, excision-repair enzymes, operate on the basis of damage or mutilation occurring to only one of the two strands of the DNA. The undamaged strand is used as a template to repair the damaged one. The excision repair of oxidized bases involves two

DNA glycosylases, Ogg1p and Ntg2p to remove the damaged bases, like 7,8-dihydro-8-oxo-guanine, 2,6-diamino-4-hydroxy-5-n-methylformamidopyrimidine, thymine glycol, and 5-hydroxycytosine (reviewed in 65). Lipid peroxides or damaged lipids are metabolized by peroxidases or lipases. Overall, antioxidant defenses seems to be approximately balanced with the generation of ROS *in vivo*. There appears to be no great reserve of antioxidant defenses in mammals, but as previously mentioned, some oxygen-derived species perform useful metabolic roles [66]. The production of H_2O_2 by activated phagocytes is the classic example of the deliberate metabolic generation of ROS for organism's advantage [67].

4.3. Exogenous Antioxidant Defenses: Compounds Derived from the Diet

The intake of exogenous antioxidants from fruit and vegetables is important in preventing the oxidative stress and cellular damage. Natural antioxidants like vitamin C and E, carotenoids and polyphenols are generally considered as beneficial components of fruits and vegetables. Their antioxidative properties are often claimed to be responsible for the protective effects of these food components against cardiovascular diseases, certain forms of cancers, photosensitivity diseases and aging [68]. However, many of the reported health claims are based on epidemiological studies in which specific diets were associated with reduced risks for specific forms of cancer and cardiovascular diseases.The identification of the actual ingredient in a specific diet responsible for the beneficial health effect remains an important bottleneck for translating observational epidemiology to the development of functional food ingredients. When ingesting high amounts of synthetic antoxidants, toxic pro-oxidant actions may be important to consider [68].

4.4. Adaptive responses and hormesis

The adaptive response is a phenomenon in which exposure to minimal stress results in increased resistance to higher levels of the same stressor or other stressors. Stressors can induce cell repair mechanisms, temporary adaptation to the same or other stressor, induce autophagy or trigger cell death [69]. The molecular mechanisms of adaptation to stress is the least investigated of the stress responses described above. It may inactivate the activation of apoptosis through caspase-9, i.e. through the intrinsic pathway, one of the main apoptotic pathways [70, 117]. Early stress responses result also in the post-translational activation of pre-existing defenses, as well as activation of signal transduction pathways that initiate late responses, namely the *de novo* synthesis of stress proteins and antioxidant defenses [65]. Hormesis is characterized by dose-response relationships displaying low-dose stimulation and high-dose inhibition [71]. Hormesis is observed also upon the exposure to low dose of a toxin, which may increase cell's tolerance for greater toxicity [35]. Reactive oxygen species (ROS) can be thought of as hormetic compounds. They are beneficial in moderate amounts and harmful in the amounts that cause the oxidative stress. Many studies investigated the induction of adaptive response by oxidative stress [72, 73, 74, 75]. An oxidative stress response is triggered when cells sense an increase of ROS, which may result from exposure of cells to low concentrations of oxidants, increased production of ROS or a decrease in antioxidant defenses. In order to survive, the cells induce the antioxidant defenses and other protective factors, such as stress proteins. Finkel and Holbrook [35] stated that the best strategy to enhance endogenous antioxidant levels may be the oxidative stress itself, based on the classical physiological concept of hormesis.

The enzymatic, non-enzymatic and indirect antioxidant defense systems could be involved in the induction of adaptive response to oxidative stress [76, 77, 78, 79, 80, 81]. It was observed, that a wide variety of stressors, such as pro-oxidants, aldehydes, caloric restriction, irradiation, UV-radiation, osmotic stress, heat shock, hypergravity, etc. can have a life-prolonging effect.The effects of these stresses are linked also to changes in intracellular redox potential, which are transmitted to changes in activity of numerous enzymes and pathways. The main physiological benefit of adaptive response is to protect the cells and organisms from moderate doses of a toxic agent [82, 69]. As such, the stress responses that result in enhanced defense and repair and even cross protection against multiple stressors could have clinical or public-health use.

4.5. Sequestration of metal ions; Fenton-like reactions

Many metal ions are necessary for normal metabolism, however they may represent a health risk when present in higher concentrations. Increased ROS generation has been implicated as a consequence of exposure to high levels of metal ions, like, iron, copper, lead, cobalt, mercury, nickel, chromium, selenium and arsenic, but not to manganese and zinc. The above mentioned transition metal ions are redox active: reduced forms of redox active metal ions participate in already discussed Fenton reaction where hydroxyl radical is generated from hydrogen peroxide [83]. Furthermore, the Haber-Weiss reaction, which involves the oxidized forms of redox active metal ions and superoxide anion, generates the reduced form of metal ion, which can be coupled to Fenton reaction to generate hydroxyl radical [15].

Fenton reaction

$$\text{Metal}^{(n+1)} + H_2O_2 \rightarrow \text{Metal}^{(n+1)+} + HO^{\cdot} + OH- \tag{5}$$

Haber-Weiss reaction

$$\text{Metal}^{(n+1)+} + 2O_2^{-\cdot} \rightarrow \text{Metal}^{(n+1)} + O_2 \tag{6}$$

Redox cycling is a characteristic of transition metals [84], and Fenton-like production of ROS appear to be involved in iron-, copper-, chromium-, and vanadium-mediated tissue damage [85]. Increases in levels of superoxide anion, hydrogen peroxide or the redox active metal ions are likely to lead to the formation of high levels of hydroxyl radical by the chemical mechanisms listed above. Therefore, the valence state and bioavailability of redox active metal ions contribute significantly to the generation of reactive oxygen species.

- The consequence of formation of free radicals mediated by metals are modifications of DNA bases, enhanced lipid peroxidation, and altered calcium and sulfhydryl homeostasis. Lipid peroxides, formed by the attack of radicals on polyunsaturated fatty acid residues of phospholipids, can further react with redox metals finally producing mutagenic and carcinogenic malondialdehyde, 4-hydroxynonenal and other exocyclic DNA adducts (etheno and/or propano adducts). The unifying factor in determining toxicity and carcinogenicity for all these metals is the abitliy to generate reactive oxygen and nitrogen species. Common mechanisms involving the Fenton reaction, generation of the superoxide radical and the hydroxyl radical are primarily associated with mitochondria, microsomes and peroxisomes. Enzymatic and non-enzymatic antioxidants protect against deleterious metal-mediated free radical attacks to some extent; e.g., vitamin E and melatonin can prevent the majority of metal-mediated (iron, copper, cadmium) damage both in *in vitro* systems and in metal-loaded animals [86, 87].

Iron Chelators

A chelator is a molecule that has the ability to bind to metal ions, e.g. iron molecules, in order to remove heavy metals from the body. According to Halliwell and Gutteridge [22] chelators act by multiple mechanisms; mainly to i) alter the reduction potential or accessibility of metal ions to stop them catalysing $OH^{.}$ production (e.g. transferrin or lactoferrin) ii) prevent the escape of the free radical into solution (e.g. albumin). In this case the free radicals are formed at the biding site of the metal ions to chelating agent. Chelators can be manmade or be produced naturally, e.g. plant phenols. Because the iron catalyzes ROS generation, sequestering iron by chelating agents is thought to be an effective approach toward preventing intracellular oxidative damage. Many chelating agents have been used to inhibit iron- or copper-mediated ROS formation, such as ethylenediaminetetraacetic acid (EDTA), diethylenetriaminepenta-acetic acid (DETAPAC), N,n'-Bis- (2-Hydroxybenzyl)ethylenediamine-N,n'-diacetic acid (HBED), 2-3-Dihydroxybenzoate, Desferrioxamine B (DFO), deferasirox (ICL 670), N,N'-bis-(3,4,5-trimethoxybenzyl) ethylenediamine N,N,-diacetic acid dihydrochloride (OR10141), phytic acid, PYSer and others (for details see 22).

Desferrioxamine can react directly with several ROS and is used as iron(III) chelator for prevention and treatment of iron overload in patients who ingested toxic oral doses of iron [22]. Also, the intracellular protein ferritin plays a role in cellular antioxidant defense. It binds nonmetabolized intracellular iron, therefore, aids to regulation of iron availability. In this way it can decrease the availability of iron for participation in Fenton reaction and lipid peroxidations. Body iron burden can be assessed by using a variety of measurements, such as serum ferritin levels and liver iron concentration by liver biopsies [for detailed information see 88, 89, 90].

4.6. Stabilizing mitochondrial ROS production

Oxidative stress and oxidative damage accumulation could be decreased by regulating the electron leakage from electron transport chain and the resultant ROS production [44]. Nutritional and lifestyle modifications may decrease mitochondrial ROS formation, e.g. by caloric restriction (CR), sport activities and healthy eating habits. The anti-aging action of caloric restriction is an example of hormesis [91, 92, 93]. The works of Yu and Lee [94], Koizumi et al. [95] and Chen and Lovry [96] imply that food restriction (energetic stress) increases the overall antioxidant capacity to maintain the optimal status of intracellular environment by balancing ROS in CR thus promotes the metabolic shift to result in more efficient electron transport at the mitochondrial respiratory chain [97]. In this way, the leakage of electrons from the respiratory chain is reduced [98, 99]. There are reports of slower aging by intermittent fasting without the overall reduction of caloric intake [100, 101]. Since it is extremely hard to maintain the long-term CR, the search is on for CR mimetics. These are the agents or strategies that can mimic the beneficial health-promoting and anti-aging effects of CR. Several compounds have been tested for a potential to act as CR mimetic; such as plant-derived polyphenols (e.g., resveratrol, quercetin, butein, piceatannol), insulin-action enhancers (e.g., metformin), or pharmacological agents that inhibit glycolysis (e.g., 2-deoxyglucose) [102].

Mitochondrial uncoupling has been proposed as a mechanism that reduces the production of reactive oxygen species and may account for the paradox between longevity and activity [103]. Moderate and regular exercise enhances health and longevity relative to sedentary lifestyles. Endurance training adaptation results in increased efficiency in ATP synthesis at the expense of potential increase in oxidative stress that is likely to be compensated by enhanced activities of antioxidant enzymes [104] and proteasome [105]. Exercise requires a large flux of energy and a shift in substrate metabolism in mitochondria from state 4 to state 3. This shift may cause an increase in superoxide production [106]. Indeed, a single bout of exercise was found to increase the metabolism and oxidative stress during and immediately after exercise [107, 108, 109]. While a single bout of exercise of sedentary animals is likely to cause increased detrimental oxidative modification of proteins [110], moderate daily exercise appears to be beneficial by reducing the damage in rat skeletal muscle [105]. Organisms exposed to oxidative stress often decrease their rate of metabolism [111, 112]. Metabolic uncoupling may reduce the mitochondrial oxidant production [113]. It may account for the paradox between longevity and activity [103]. Heat is produced when oxygen consumption is uncoupled from ATP generation. When the mitochondria are uncoupled and membrane potential is low animals might produce less free radicals when expending the most energy [114]. Postprandial oxidative stress is characterized by an increased susceptibility of the organism toward oxidative damage after consumption of a meal rich in lipids and/or carbohydrates [115]. The generation of excess superoxide due to abundance of energy substrates after the meal may be a predominate factor resulting in oxidative stress and a decrease in nitric oxide. A mixture of antioxidant compounds is required to provide protection from the oxidative effects of postprandial fats and sugars. No specific antioxidant can be claimed to be the most important, as consumption of food varies enormously in humans. However, a variety of polyphenolic compounds derived from plants appear to be effective dietary antioxidants, especially when consumed with high-fat meals [116].

5. Conclusion and perspectives

In conclusion, excessive production of ROS and reduced antioxidant defence with age significantly contribute to aging. It seems that oxidative damage is the major cause and the most important contributor to human aging. Antioxidant defense seems to be approximatly balanced with the generation of oxygen-derived species in young individuals, however, there is an increase of oxidative stress later in life. Then the approaches to lower the increased ROS formation in our bodies could be implemented by avoiding the exposure to exogenous free radicals, by intake of adequate amounts of antioxidants and/or by stimulating the damage-repair systems of the cells [44 and references within].

Developing natural or pharmacological agents capable of increasing the antioxidative protection and/or modulating the endogenous defense and repair mechanisms may potentially improve health, increase longevity and contribute to treatment of degenerative age-related diseases, such as cardiovascular and neurodegenerative disorders and cancer. The lifestyle changes, e.g. regular physical activity, increased intake of fruits and vegetables, and reduced calorie intake may improve health and increase cellular resistance to stress. Synthetic antioxidant supplements may help to correct the high levels of oxidative stress that cannot be controlled by the sinergy of endogenous antioxidant systems.

Author details

B. Poljsak[1*] and I. Milisav[1,2]

*Address all correspondence to: borut.poljsak@zf.uni-lj.si

1 University of Ljubljana, Laboratory of oxidative stress research, Faculty of Health Sciences, Ljubljana, Slovenia

2 University of Ljubljana, Faculty of Medicine, Institute of Pathophysiology, Ljubljana, Slovenia

References

[1] Kirkwood, B., & Mathers, J. C. (2009). The basic biology of aging. In: Stanner S., Thompson R., Buttriss J. [eds.], *Healthy aging-The role of nutrition and lifestyle*, NY:Wiley-Blackwell.

[2] Fontana, L., & Klein, S. (2007). Aging, Adiposity, and Calorie Restriction. *Jama*, 297, 986-994.

[3] Davies, K. J. (1995). Oxidative stress: the paradox of aerobic life. *Biochem. Soc. Symp.*, 61, 1-31.

[4] Martin, G. M. (1987). Interaction of aging and environmental agents: The gerontological perspective. *Prog. Clin. Bio. Res*, 228, 25-80.

[5] Martin, G. M., Austad, S. N., & Johnson, T. E. (1996). Genetic analysis of ageing: Role of oxidative damage and environmental stress. *Nat. Genet*, 13, 25-34.

[6] Gilca, M., Stoian, I., Atanasiu, V., & Virgolici, B. (2007). The oxidative hypothesis of senescence. *J. Postgrad. Med.*, 53(3), 207-213.

[7] Beckman, K. B., & Ames, B. N. (1998). The free radical theory of aging matures. *Physiol. Rev.*, 78, 547-581.

[8] Casteilla, L., Rigoulet, M., & Penicaud, L. (2001). Mitochondrial ROS metabolism: Modulation by uncoupling proteins. *IUBMB Life*, 52, 181-188.

[9] Hansford, R. G., Hogue, B. A., & Mildaziene, V. (1997). Dependence of H2O2 formation by rat heart mitochondria on substrate availability and donor age. *J. Bioenerg. Biomembr.*, 29, 89-95.

[10] Staniek, K., & Nohl, H. (1999). H(2)O(2) detection from intact mitochondria as a measure for one-electron reduction of dioxygen requires a non-invasive assay system. *Biochim Biophys Acta.*, 1413(2), 70-80.

[11] Speakman, J. R., Selman, C., McLaren, J. S., & Harper, E. J. (2002). Living fast, dying when? The link between aging and energetics. *J. Nutr.*, 132, 1583S-97S.

[12] Mooijaart, S. P., van Heemst, D., Schreuder, J., van Gerwen, S., Beekman, M., & Brandt, B. W. (2004). Variation in the SHC1 gene and longevity in humans. *Exp. Gerontol*, 39, 263-8.

[13] Harman, D. (1972). A biologic clock: the mitochondria? *Journal of the American Geriatrics Society*, 20, 145-147.

[14] Harman, D. (1956). Aging: a theory based on free radical and radiation chemistry. *Journal of Gerontology*, 11, 298-300.

[15] Halliwell, B., & Gutteridge, J. (1999). Free radicals in biology and medicine [3rd edn]. *Oxford: Clarendon Press.*

[16] Reiter, R. J. (1995). Oxygen radical detoxification processes during aging: The functional importance of melatonin. *Aging (Milano)*, 7, 340-51.

[17] Hagen, J. L., Krause, D. J., Baker, D. J., Fu, M. H., Tarnopolsky, M. A., & Hepple, R. T. (2004). Skeletal muscle aging in F344BN F1-hybrid rats: I. mitochondrial dysfunction contributes to the age-associated reduction in CO2max. *J. Gerontol. A. Biol. Sci. Med. Sci.*, 59, 1099-1110.

[18] Hamilton, M. L., Van Remmen, H., Drake, J. A., Yang, H., Guo, Z. M., Kewitt, K., Walter, C. A., & Richardson, A. (2001). Does oxidative damage to DNA increase with age? *Proc. Natl. Acad. Sci. USA*, 98, 10469-10474.

[19] Hagen, T. M. (2003). Oxidative stress, redox imbalance, and the aging process. *Antioxid. Redox Signal*, 5, 503-506.

[20] Sohal, R. (2002). Role of oxidative stress and protein oxidation in the aging process. *Free Radic Biol. Med.*, 33, 37-44.

[21] Sohal, R., Mockett, R., & Orr, W. (2002). Mechanisms of aging: an appraisal of the oxidative stress hypothesis. *Free Radic. Biol. Med.*, 33, 575-86.

[22] Halliwell, B., & Gutteridge, J. (2007). Free radicals in biology and medicine [4th edn]. *Oxford: University Press*.

[23] Ames, B. N. (2004). A Role for Supplements in Optimizing Health: the Metabolic Tune-up. *Archives of Biochemistry and Biophysics*, 423, 227-234.

[24] Arnheim, N., & Cortopassi, G. (1992). Deleterious mitochondrial DNA mutations accumulate in aging human tissues. *Mutat Res.*, 275(3-6), 157-67.

[25] Yang, J. H., Lee, H. C., Lin, K. J., & Wei, Y. H. (1994). A specific 4977- bp deletion of mitochondrial DNA in human aging skin. *Arch. Dermatol. Res*, 286, 386-390.

[26] Golden, T., Morten, K., Johnson, F., Samper, E., & Melov, S. (2006). Mitochondria: A critical role in aging. In: Masoro EJ., Austad S. [eds.], *Handbook of the biology of aging*, Sixth edition. Elsevier.

[27] Jacobs, H. T. (2003). The mitochondrial theory of aging: dead or alive? *Aging cell*, 2, 11.

[28] Pak, J. W., Herbst, A., Bua, E., Gokey, N., McKenzie, D., & Aiken, J. M. (2003). Rebuttal to Jacobs: the mitochondrial theory of aging: alive or dead. *Aging Cell*, 2, 9.

[29] De Grey, A. D. N. J. (2005). Reactive Oxygen Species Production in the Mitochondrial Matrix: Implications for the Mechanism of Mitochondrial Mutation Accumulation. *Rejuvenation Res.*, 8(1), 13-7.

[30] Best, B. Mechanisms of Aging. http://www.benbest.com/lifeext/aging.html, [accessed 10 May 2012].

[31] Chung, H. Y., Sung, B., Jung, K. J., Zou, Y., & Yu, B. P. (2006). The molecular inflammatory process in aging. *Antioxid. Redox. Signal.*, 8, 572-581.

[32] Navratil, V. (2011). Health, Ageing and Entropy. *School and Health 21 Health Literacy Through Education.*, Vyd. 1. Brno : Masarykova Univerzita, 978-8-02105-720-3, 329-336, Brno.

[33] Terman, A. (2001). Garbage catastrophe theory of aging: Imperfect removal of oxidative damage? *Redox. Rep.*, 6, 15-26.

[34] Schulz, T. J., Zarse, K., Voigt, A., Urban, N., Birringer, M., & Ristow, M. (2007). Glucose Restriction Extends Caenorhabditis elegans Lifespan by Inducing Mitochondrial Respiration and Increasing Oxidative Stress. *Cell Metabolism*, 6, 280-293.

[35] Finkel, T., & Holbrook, Nikki J. (2000). Oxidants, oxidative stress and the biology of ageing. *Nature*, 408, 239-247.

[36] Kirkwood, T. B. L., & Holliday, R. (1979). The evolution of ageing and longevity. *Proc. R. Soc. London Ser. B Biol. Sci.*, 205, 531-546.

[37] Kirkwood, T. B. L. (1997). Evolution of ageing. *Nature*, 270, 301-304.

[38] Gavrilov, L. A., & Gavrilova, N. S. (2002). Evolutionary Theories of Aging and Longevity. *The Scientific World JOURNAL*, 2, 339-356.

[39] Von Zglinicki, T., Bürkle, A., & Kirkwood, T. B. (2001). Stress, DNA damage and ageing-an integrative approach. *Exp. Gerontol.*, 36, 1049-1062.

[40] Kurz, D. J., Decary, S., Hong, Y., Trivier, E., Akhmedov, A., & Erusalimsky, J. D. (2004). Chronic oxidative stress compromises telomere integrity and accelerates the onset of senescence in human endothelial cells. *J. Cell Sci*, 117, 2417-2426.

[41] Bayne, S., & Liu, J. P. (2005). Hormones and growth factors regulate telomerase activity in aging and cancer. *Mol. Cell Endocrinol.*, 240, 11-22.

[42] Arking, R. (2006). The biology of aging, observations and principles. *Third edition. New York: Oxford University Press.*

[43] Poljsak, B. (2011). Skin aging, free radicals and antioxidants. *New York: NovaScience Publisher.*

[44] Poljsak, B. (2011). Strategies for reducing or preventing the generation of oxidative stress. *Oxid Med Cell Longev.*, 194586.

[45] Cheeseman, K. H., & Slater, T. F. (1993). An introduction to free radical biochemistry. *Br. Med. Bull*, 49, 481-493.

[46] Hohmann, S. (1997). Yeast Stress Responses. In Hohmann S., Mager WH [eds], *Yeast stress responses*, Austin: HRG. Landes Company.

[47] Santoro, N., & Thiele, D. J. (1997). Oxidative stress responses in the yeast Saccharomyces cerevisiae. In Hohmann S., Mager WH [eds], *Yeast stress responses*, Austin: HRG. Landes Company.

[48] Chesney, J. A., Eaton, J. W., & Mahoney, J. (1996). Bacterial glutathione: a sacrificial defense against chlorine compounds. *J. Bacteriol.*, 178(7), 2131-2135.

[49] Jamnik, P., & Raspor, P. (2003). Stress response of yeast Candida intermedia to Cr(VI). *J. Biochem. Mol. Toxicol.*, 17, 316-23.

[50] Aiyar, J., Berkovits, H. J., Floyd, R. A., & Wetterhahn, K. E. (1990). Reaction of chromium (VI) with hydrogen peroxide in the presence of glutathione: reactive intermediates and resulting DNA damage. *Chem Res Toxicol*, 3(6), 595-603.

[51] Wetterhahn, K. E., & Hamilton, J. W. (1989). Molecular basis of hexavalent chromium carcinogenicity: effect on gene expression. *Sci Total Environ*, 86(1-2), 113-29.

[52] Izawa, S., Inoue, Y., & Kimura, A. (1995). Oxidative stress response in yeast: effect of glutathione on adaptation to hydrogen peroxide stress in Saccharomyces cerevisiae. *Fabs Lett*, 368, 73-76.

[53] Terman, A., & Brunk, U. T. (2006). Oxidative stress, accumulation of biological "garbage," and aging. *Antioxid. Redox Signal*, 8, 197-204.

[54] Brunk, U. T., Jones, C. B., & Sohal, R. S. (1992). A novel hypothesis of lipofuscinogenesis and cellular aging based on interaction between oxidative stress and autophagocitosis. *Mutat. Res*, 275, 395-403.

[55] Cortopassi, G. A., & Wang, E. (1996). There is substantial agreement among interspecies estimates of DNA repair activity. *Mechanisms of Aging and Development*, 91, 211-218.

[56] Grune, T., Reinheckel, T., & Davies, K. J. (1997). Degradation of oxidized proteins in mammalian cells. *Faseb. J.*, 11, 526-34.

[57] Chondrogianni, N., Kapeta, S., Chinou, I., Vassilatou, K., Papassideri, I., & Gonos, E. S. (2010). Anti-ageing and rejuvenating effects of quercetin. *Exp. Gerontol.*, 45(10), 763-71.

[58] Stadtman, E. R. (1992). Protein oxidation and aging. *Science*, 257, 1220-4.

[59] Shringarpure, R., & Davies, K. J. (2002). Protein turnover by the proteasome in aging and disease. *Free Radic. Biol. Med.*, 32, 1084-9.

[60] Sell, D. R., Lane, M. A., Johnson, W. A., Masoro, E. J., Mock, O. B., Reiser, K. M., Fogarty, J. F., Cutler, R. G., Ingram, D. K., Roth, G. S., & Monnier, V. M. (1996). Longevity and the genetic determination of collagen glycoxidation kinetics in mammalian senescence. *Proc. Natl. Acad. Sci. USA*, 93(1), 485-90.

[61] Torres, C. A., & Perez, V. I. (2008). Proteasome modulates mitochondrial function during cellular senescence. *Free Radic. Biol. Med.*, 44(3), 403-14.

[62] Promislow, D. E. (1994). DNA repair and the evolution of longevity: a critical analysis. *J.Theor. Biol.*, 170, 291-300.

[63] Bürkle, A., Beneke, S., Brabeck, C., Leake, A., Meyer, R., Muiras, M. L., & Pfeiffer, R. (2002). Poly(ADP-ribose) polymerase-1, DNA repair and mammalian longevity. *Exp. Gerontol.*, 37(10-11), 1203-5.

[64] Grube, K., & Bürkle, A. (1992). Poly(ADP-ribose) polymerase activity in mononuclear leukocytes of 13 mammalian species correlates with species-specific lifespan. *Proc. Natl. Acad. Sci. USA*, 89, 11759-11763.

[65] Costa, V., & Moradas-Ferreira, P. (2001). Oxidative stress and signal transduction in Saccharomyces cerevisiae: insights into ageing, apoptosis and diseases. *Mol. Aspects Med.*, 22, 217-246.

[66] Stocker, R., & Frei, B. (1991). Endogenous antioxidant defenses in human blood plasma. *In: Oxidative stress: oxidants and antioxidants.*, London: Academic press.

[67] Halliwell, B., & Cross, C. E. (1994). Oxygen-derived species: their role in human disease and environmental stress. *Environ. Health Perspect.*, 102, 5-12.

[68] Rietjens, I., Boersma, M., & de Haan, L. (2001). The pro-oxidant chemistry of the natural antioxidants vitamin C, vitamin E, carotenoids and flavonoids. *Environ Toxicol. Pharmacol*, 11, 321-333.

[69] Milisav, I. (2011). Cellular Stress Responses. In: Wislet-Gendebien S. [Ed.], *Advances in Regenerative Medicine*, 978-9-53307-732-1, InTech, Available from, http://www.intechopen.com/articles/show/title/cellular-stress-responses.

[70] Nipic, D., Pirc, A., Banic, B., Suput, D., & Milisav, I. (2010). Preapoptotic cell stress response of primary hepatocytes. *Hepatology.*, 51(6), 2140-51.

[71] Calabrese, E. J., & Baldwin, L. A. (2002). Hormesis and high-risk groups. *Regul Toxicol Pharmacol.*, 35(3), 414-28.

[72] Feinendegen, L. E., Bond, V. P., Sondhaus, C. A., & Muehlensiepen, H. (1996). Radiation effects induced by low doses in complex tissue and their relation to cellular adaptive responses. *Mutat Res*, 358, 199-205.

[73] Jones, S. A., McArdle, F., Jack, C. I. A., & Jackson, M. J. (1999). Effect of antioxidant supplement on the adaptive response of human skin fibroblasts to UV-induced oxidative stress. *Redox Report*, 4, 291-299.

[74] de Saint-Georges, L. (2004). Low-dose ionizing radiation exposure: Understanding the risk for cellular transformation. *J Biol Regul Homeost Agents*, 18, 96-100.

[75] Shankar, B., Pandey, R., & Sainis, K. (2006). Radiation-induced bystander effects and adaptive response in murine lymphocytes. *Int J Radiat Biol*, 82, 537-548.

[76] Mendez-Alvarez, S., Leisinger, U., & Eggen, R. I. (1999). Adaptive responses in Chlamydomonas reinhardtii. *Int Microbiol*, 2, 15-22.

[77] Chen, Z. H., Yoshida, Y., Saito, Y., Sekine, A., Noguchi, N., & Niki, E. (2006). Induction of adaptive response and enhancement of PC12 cell tolerance by 7-hydroxycholesterol and 15-deoxy-delta(12,14)-prostaglandin J2 through up-regulation of cellular glutathione via different mechanisms. *J Biol Chem*, 281, 14440-14445.

[78] Yan, G., Hua, Z., Du, G., & Chen, J. (2006). Adaptive response of Bacillus sp. F26 to hydrogen peroxide and menadione. *Curr Microbiol*, 52, 238-242.

[79] Tosello, M. E., Biasoli, M. S., Luque, A. G., Magaró, H. M., & Krapp, A. R. (2007). Oxidative stress response involving induction of protective enzymes in Candida dubliniensis. *Med Mycol*, 45, 535-540.

[80] Joksic, G., Pajovic, S. B., Stankovic, M., Pejic, S., Kasapovic, J., Cuttone, G., Calonghi, N., Masotti, L., & Kanazir, D. T. (2000). Chromosome aberrations, micronuclei, and activity of superoxide dismutases in human lymphocytes after irradiation in vitro. *Cell Mol Life Sci*, 57, 842-850.

[81] Bercht, M., Flohr-Beckhaus, C., Osterod, M., Rünger, T. M., Radicella, J. P., & Epe, B. (2007). Is the repair of oxidative DNA base modifications inducible by a preceding DNA damage induction? *DNA Repair*, 6, 367-373.

[82] Crawford, D. R., & Davies, K. J. (1994). Adaptive response and oxidative stress. *Environ Health Perspect.*, 102(10), 25-8.

[83] Nordberg, J., & Arner, E. S. J. (2001). Reactive oxygen species, antioxidants, and the mammalian thioredoxin system. *Free Radic Biol Med*, 31(11), 1287-1312.

[84] Klein, C. B., Frenkel, K., & Costa, M. (1991). The role of oxidative processes in metal carcinogenesis. *Chem. Res. Toxicol.*, 4, 592-604.

[85] Fuch, J., Podda, M., & Zollner, T. (2001). Redox Modulation and Oxidative Stress in Dermatotoxicology. *In: Fuchs, J; Packer, L. [eds]. Environmental stressors in health and disease. NY: Marcel Dekker, Inc.*

[86] Valko, M., Morris, H., & Cronin, M. T. (2005). Metals, toxicity and oxidative stress. *Curr. Med. Chem.*, 12(10), 1161-208.

[87] Valko, M., Leibfritz, D., Moncol, J., Cronin, M. T. D., Mazur, M., & Telser, J. (2007). Free radicals and antioxidants in normal physiological functions and human disease. *The International Journal of Biochemistry & Cell Biology*, 39, 44-84.

[88] Jensen, P. D. (2004). Evaluation of iron overload. *Br J Haematol*, 124(6), 697-71.

[89] Angelucci, E., Brittenham, G. M., McLaren, C. E., et al. (2000). Hepatic iron concentration and total body iron stores in thalassemia major. *N Engl J Med.*, 343(5), 327-331.

[90] Kitazawa, M., Iwasaki, K., & Sakamoto, K. (2006). Iron chelators may help prevent photoaging. *J. Cosmet. Dermatol.*, 5(3), 210-7.

[91] Anderson, R. M., Bitterman, K. J., Wood, J. G., Medvedik, O., & Sinclair, D. A. (2003). Nicotinamide and Pnc 1 govern lifespan extension by calorie restriction in S. *Cerevisiae. Nature*, 432, 181-185.

[92] Iwasaki, K., Gleiser, C. A., Masoro, E. J., McMahan, C. A., Seo, E. J., & Yu, B. P. (1988). The influence of the dietary protein source on longevity and age-related disease processes of Fischer rats. *Journal of gerontology*, 43, B5-B12.

[93] Mattson, M. P. (2003). Energy Metabolism and Lifespan Determination. *Adv. Cell Aging Geronto*, 14, 105-122.

[94] Lee, D. W., & Yu, B. P. (1991). Food restriction as an effective modulator of free radical metabolism in rats. *Korean Biochem J*, 24, 148-154.

[95] Koizumi, A., Weindruch, R., & Walford, R. L. (1987). Influences of dietary restriction and age on liver enzyme activities and lipid peroxidation in mice. *J Nutr*, 117(2), 361-7.

[96] Chen, L. H., & Lowry, S. R. (1989). Cellular antioxidant defense system. *Prog Clin Biol Res*, 287, 247-56.

[97] Sohal, R., & Weindruch, R. (1996). Oxidative stress, caloric restriction, and aging. *Science*, 273, 59-63.

[98] Korshunov, S. S., Skulachev, V. P., & Starkov, A. A. (1997). High protonic potential actuates a mechanism of production of reactive oxygen species in mitochondria. *Febs. Lett.*, 416, 15-18.

[99] Starkov, A. A. (1997). "Mild" uncoupling of mitochondria. *Biosci. Rep.*, 17, 273-279.

[100] Gredilla, R., Sanz, A., Lopez-Torres, M., & Barja, G. (2001). Caloric restriction decreases mitochondrial free radical generation at complex I and lowers oxidative damage to mitochondrial DNA in the rat heart. *Faseb J.*, 15, 1589-1591.

[101] Anson, R. M., Guo, Z., de Cabo, R., Iyun, T., Rios, M., Hagepanos, A., Ingram, D. K., Lane, M. A., & Mattson, M. P. Intermittent fasting dissociates beneficial effects of dietary restriction on glucose metabolism and neuronal resistance to injury from calorie intake. *Proc. Natl. Acad. Sci. U S A.*, USA, 203, 100(10), 6216-20.

[102] Ingram, D. K., Zhu, M., & Mamczarz, J. (2006). Calorie restriction mimetics: an emerging research field. *Aging Cell.*, 5, 97-108.

[103] Cámara, Y., Duval, C., Sibille, B., & Villarroya, F. (2007). Activation of mitochondrial-driven apoptosis in skeletal muscle cells is not mediated by reactive oxygen species production. *Int. J. Biochem. Cell Biol.*, 39(1), 146-60.

[104] Hollander, J., Fiebig, R., Gore, M., Bejma, J., Ookawara, T., Ohno, H., & Ji, L. L. (1999). Superoxide dismutase gene expression in skeletal muscle: fiber-specific adaptation to endurance training. *Am. J. Physiol.*, 277, R856-R862.

[105] Radak, Z., Nakamura, A., & Nakamoto, H. (1998). A period of exercise increases the accumulation of reactive carbonyl derivatives in the lungs of rats. *Pfluger Arch: Eur. J. Physiol.*, 435, 439-441.

[106] Barja, G. (1999). Mitochondrial oxygen radical generation and leak: sites of production in states 4 and 3, organ specificity, and relation to aging and longevity. *J Bioenerg Biomembr.*, 31(4), 347-66.

[107] Alessio, H. M., & Goldfarb, A. H. (1988). Lipid peroxidation and scavenger enzymes during exercise. Adaptive response to training. *J Appl Physiol*, 64, 1333-1336.

[108] Ji, L. L. (1993). Antioxidant enzyme response to exercise and aging. *Med Sci Sport Exerc.*, 25, 225-231.

[109] Powers, S. K., & Jackson, M. J. (2008). Exercise-induced oxidative stress: cellular mechanisms and impact on muscle force production. *Physiol Rev.*, 88(4), 1243-76.

[110] Reznick, A. Z., Kagan, V. E., Ramsey, R., Tsuchiya, M., Khwaja, S., Serbinova, E. A., & Packer, L. (1992). Antiradical effects in L-propionyl carnitine protection of the heart against ischemia-reperfusion injury: the possible role of iron chelation. *Arch Biochem Biophys.*, 296(2), 394-401.

[111] Allen, R. G., Farmer, K. J., Newton, R. K., & Sohal, R. S. (1984). Effects of paraquat administration on longevity, oxygen consumption, lipid peroxidation, superoxide dismutase, catalase, glutathione reductase, inorganic peroxides and glutathione in the adult housefly. *Comp Biochem Physiol C.*, 78(2), 283-8.

[112] Allen, R. G., & Sohal, R. S. (1982). Life-lengthening effects of gamma-radiation on the adult housefly, Musca domestica. *Mech Ageing Dev.*, 20(4), 369-75.

[113] Skulachev, V. P. (1996). Role of uncoupled and non-coupled oxidations in maintenance of safely low levels of oxygen and its one-electron reductants. *Q Rev Biophys.*, 169-202.

[114] Speakman, J. R., & Selman, C. (2011). The free-radical damage theory: Accumulating evidence against a simple link of oxidative stress to ageing and lifespan. *Bioessays.*, 33(4), 255-9.

[115] Ursini, F., & Sevanian, A. (2002). Postprandial oxidative stress. *Biol Chem.*, 383(3-4), 599-605.

[116] Sies, H., Stahl, W., & Sevanian, A. (2005). Nutritional, dietary and postprandial oxidative stress. *J Nutr.*, 135(5), 969-72.

[117] Banič, B., Nipič, D., Suput, D., & Milisav, I. (2011). DMSO modulates the pathway of apoptosis triggering. *Cell Mol Biol Lett.*, 16(2), 328-41.

Disease and Therapy - A Role for Antioxidants

Emerging Role of Natural Antioxidants in Chronic Disease Prevention with an Emphasis on Vitamin E and Selenium

Manuel Soriano García

Additional information is available at the end of the chapter

1. Introduction

The possibility has arisen within the last three decades that major diseases that directly affect humankind worldwide may be preventable by the simple improving the dietary intake of those nutrient substances that have become called "antioxidant nutrients".

There is no doubt that successful prevention is the key to controlling morbidity and mortality from chronic diseases affecting humankind. Prevention provides: the methods to avoid occurrence of disease and most population-based health promotion efforts are of this type; methods to diagnose and treat extant disease in early stages before it causes significant morbidity; methods to reduce negative impact of extant disease by restoring function and reducing disease-related complications; and finally, the methods to mitigate or avoid results of unnecessary or excessive interventions in the health system.

The quality and quantity of diet with respect to the intake of fresh food (fruits, seeds and vegetables) may improve our health and consequently decrease the risk of any disease. Currently, the antioxidant nutrients are the vitamins C and E and β-carotene. However, it is worthy to mention that these compounds are involved in other functions a part from being antioxidant nutrients.

Selenium (Se), a trace mineral. Is the 34th element and is located between sulfur and tellurium in Group 16 in the periodic table. It is a nonmetallic element and its properties are intermediate between adjacent sulfur and tellurium. It was originally discovered by a German chemist Martin Heinric Klaproth, but misidentified as tellurium. Later, in 1818 a Swedish chemist Jons Jacob Berzelius discovered selenium and was named after the Greek goddess of the moon, Selene [1] and its name was associated with tellurium, a name for earth. He observed the element as a deposit following the oxidation of sulfur dioxide from cooper pyrites. It ranks seventieth in abundance among the elements and is distributed in the Earth's crust at concentrations averaging 0.09 mg/kg [2]. Selenium has six major stable isotopes have been reported and the most abundant in nature are 80Se (49.6%) and 78Se (23.8%) [3]. In general, selenium is present in the environment in elemental form or in the form of selenide (Se^{2-}), selenate (SeO_4^{2-}), or selenite (SeO_3^{2-}). The identity and amounts of the various oxidation-state species in soils depends enormously on the redox-potential conditions. The lower oxidation states predominate in anaerobic conditions, acidic soils, and the higher oxidation states are favored in alkaline and aerobic conditions. Both selenites and selenates are taken up by plants and converted to protein-bound selenocysteine and selenomethionine, soluble inorganic forms, several free amino acids, and volatile organoselenium compounds. The elemental form of selenium, selenium dioxide, and volatile organoselenium compounds produced by industries and plants are incorporated in the environment. Selenium occurs naturally in water in trace amounts as a result of geochemical processes, such as weathering of rocks and erosion of soils, and is usually present in water as selenate or selenite; however the elemental form may be carried in suspension [4].

Interest in selenium and health was focused primarily on the potentially toxic effects of high intakes in humans, stimulated by reports of alkali disease in livestock raised in seleniferous areas, in the last century [5]. Selenium is a trace mineral that is essential to good health but required only small amounts [6,7]. Selenium is considered as essential human micronutrient and is incorporated into proteins to make selenoproteins. Selenium is present in the selenoproteins, as the aminoacid selenocysteine (Se-Cys) [8-12].

Dietary levels of the desired amount of Se are in a very narrow range: consumption of foods containing less than 0.1 mg kg-1 of this element will result in Se deficiency, whereas dietary levels above 1 mg kg-1 will lead to toxic manifestations [13]. Se status varies significantly across different populations and different ethnic groups [14-15].

Selenium enters the food chain through plants, and the amount and bioavailability of selenium in the soil typically reflects the plant level. Selenium is provided by the diet in humans, but may also be provided from drinking water, environmental pollution, and in recent years through supplementation [16,17]. Plants convert Se mainly into selenomethionine (Se-Met) and incorporated it into protein place of methionine. More than 50% of the total Se content of the plant exist as Se-Met, the rest exist as selenocysteine (Se-Cys), methyl-Se-Cys and c-glutamyl-Se-methyl-Cys. The later compounds are not significantly incorporated into plant protein. Higher animals are unable to synthesize Se-Met and only Se-Cys was detected in rats supplemented with Se as selenite [18]. Animals that eat grains (Brazil nuts, sunflower seeds, walnuts and grains) that were grown in selenium rich soil have higher levels of selenium in their muscles, liver, kidney, heart, spleen and fingernails. Other natural selenium sources are butter, eggs, brewer's yeast, wheat germ, garlic, raspberry leaf, radish, horseradish, onions, shellfish, broccoli, fennel seed and ginseng, among other sources.

Most ingested forms of selenium ultimately are metabolized to low molecular weight inorganic and organic compounds that play a central role in human health either via incorporation into selenoproteins or binding to selenium binding proteins [19]. Therefore, a tremendous effort has been directed toward the synthesis of stable organoselenium compounds that could be used as antioxidants, enzyme modulators, antitumor, antimicrobials, antihypertensive agents, antivirals and cytokine inducers. Several excellent books and reviews appeared in literature describing the biological function of organoselenium compounds [20-22].

The role of organoselenium compounds as antioxidants, as enzyme modulators, photo-chymotherapeutic agents, cytokine inducers and immunomodulators, and antihypertensive and cardiotonic agents have been recently described in literature [23].

The essentiality of selenium results as a necessary component of the active center of a number of selenoenzymes. Selenium functions as a redox center. The term selenoprotein is any protein that includes in its primary sequence of amino acids, the selenocysteine (Se-Cys) residue [24]. There are at least 30 selenoproteins that have been identified in mammals, and it has been estimated that humans have about 25 selenoproteins, including glutathione peroxidase, thioredoxin reductase, iodothyronine, deiodinase, and selenoproteins P, W, and R [25-27]. GPx accounts for 10–30% of plasma selenium, and selenoprotein P accounts for another 50% [28]. These enzymes protect cells from free radical damage and regulate DNA transcription and cell proliferation. The glutathione and thioredoxin systems in particular have long been considered the major pathways through which selenium exerts its potential chemopreventive effect [24], while some investigations have also suggested growth inhibitory, proapoptotic activity for selenometabolites in premalignant cells [29]. Selenium is also involved in thyroid function, T cell immunity, and spermatogenesis [28], and is a competitive antagonist of potentially carcinogenic heavy metals such as arsenic and cadmium [30].

The organism has several biological defense mechanisms against intracellular oxidative stress such as superoxide dismutase, catalase, glutathione peroxidase and nonenzymatic antioxidants such as glutathione, vitamins A, C and E, riboflavin, a B vitamin and selenium can also contribute to overcome oxidative stress [31].

Vitamin E is a fat-soluble vitamin known for its antioxidant capacity that is why it is well known as a lipophilic antioxidant that protects membranes from being oxidatively damaged as an electron donor to free radicals [32]. Vitamin E belongs to a group of the eight naturally occurring vitamer forms, four tocopherols (α, β, γ, δ) and four tocotrienols (α, β, γ, δ) based on the hydroxyl and methyl substitution in their phenolic rings, all of which have saturated and three double bonds in their phytyl tails. α-tocopherol (from the Greek *tokos* = child, *phero* = to bear and *ol* indicating that the substance is an alcohol, is the most abundant form in nature; it's the most active and corrects human E deficiency symptoms [33]. The most abundant sources of vitamin E are vegetable oils, which typically contain all four tocopherol (α, β, γ, δ) in varying proportions, Other important source are nuts and seeds such as sunflower and amaranth seeds.

It is well known that all forms of vitamin E are lipid soluble they easily absorbed from the intestinal lumen after dietary intake via micelles created by biliary and pancreatic secretions [34-35]. Vitamin E is then incorporated into chylomicrons and secreted into the circulation where, transported by various lipoproteins, it travels to the liver [36]. Plasma α-tocopherol concentrations in humans range from 11 to 37 μmol/L, whereas γ-tocopherol are between 2 and 5 μmol/L. The liver plays a central role in regulating α-tocopherol levels by directly acting on the distribution, metabolism, and excretion of this vitamin [37]. The major hepatic regulatory mechanism is the α-tocopherol transfer protein, α-TTP, which has been identified in a variety of mammals, including humans [38]. This protein facilitates secretion of α-tocopherol from the liver into the bloodstream, by acquiring it from endosomes and then delivering it to the plasma membrane where it is released and promptly associates with the different nascent lipoproteins [39]. Plasma concentration of vitamin E depends completely on the absorption, tissue delivery, and excretion rate. The estimated α-tocopherol half-life in plasma of healthy individuals is ~ 48 to 60 H, which is much longer than the half-life of γ-tocopherol approximately 15 H. These kinetic data underscore an interesting concept that while α-tocopherol levels are maintained, the other forms of vitamin E are removed much more rapidly [40].

2. Selenium and Health

Selenium deficiency is associated with the pathogenesis of wide variety of processes that affects our health and disease including the antioxidant activity, depression, allergies, preventing oxidative stress, HIV infection, in the brain, thyroid metabolism, cancer, diabetes mellitus, male fertility, asthma, cardiovascular disorders, rheumatoid arthritis, pre-eclampsia, in immune function, in alleviate bone impairments, aging, gastrointestinal problems, selenium interactions and toxicity, anti-inflammatory effects, and hypertension. The list of clinical disorders expected to be influenced by Se deficiency is rapidly growing with time. Some selected issues regarding the role of Se in health and disease have been briefly outlined as follows:

2.1. Se and antioxidant activity

Selenocysteine is recognized as the 21st amino acid, and it forms a predominant residue of selenoproteins and selenoenzymes in biological tissues. The molecular structure of selenocystiene is an analogue of cysteine where a sulphur atom is replaced by Se. Even though Se and sulphur share some similar chemical properties, there are also some differences. The R-SeH with a pKa 5.2 is more is more acidic than R-SH with a pKa 8.5, and readily dissociated at physiological pH, which may contribute to its biological reactivity. In the body, both organic [selenocysteine(SeCys) and selenomethionine (SeMet)] and inorganic (selenite, selenate) Se compounds are readily metabolized to various forms of Se metabolites [41]. Of particular importance during this metabolic process is the formation of hydrogen selenide (H_2Se) from selenite after the action of glutathione-coupled reactions *via* selenodiglutathione (GS-Se-SG) and glutathione selenopersulfide (GS-SeH). H_2Se is further metabolized and involved in the formation of methylselenol and dimethylselenide, which are exhaled or secreted *via* the skin. Selenium is also excreted in urine as trimethylselenonium ion and selenosugar compounds [42]. The selenoproteins are classified on the basis of their biological function [25]. The first identified selenoprotein was glutathione peroxidase 1 (GPx1). The selenoenzymes with strong antioxidant activity are GPx, GPx1, GPx3, GPx4, GPx5 and GPx6. In Humans GPx1 through GPx4 and GPx6 are selenocysteine containing enzymes. These GPx play a significant role in protecting cells against oxidative damage from reactive oxygen species (ROS) and reactive nitrogen species (RNS), which include superoxide, hydrogen peroxide, hydroxyl radicals, nitric oxide and peroxynitrite [43-44]. The other essential antioxidant selenoenzymes are the thioredoxin reductase (TrxR) where they use thioredoxin (Trx) as a substrate to maintain a Trx/TrxR system in a reduced state for removal of harmful hydrogen peroxide and there are three types of TrxR. Iodothyronine deiodinase (DIO) have three subtypes, DIO 1, 2, and 3 [45].

2.2. Se and depression

In [46] selenium's function as an antioxidant, and as a constituent of selenoproteins that are important in redox homeostasis, warrants further investigation as a risk factor for depression, and suggest a potentially novel modifiable factor in the primary prevention and management of depression. Depression is becoming recognized as an inflammatory disorder, accompanied by an accumulation of highly reactive oxygen species that overwhelm usual defensive physiological processes [47-51]. Several indicators support a role for selenium in normal brain function. During times of selenium deficiency, there is preferential storage of selenium in the brain [52]. Selenium has significant modulatory effects on dopamine [53] and dopamine plays a role in the pathophysiology of depression and other psychiatric illnesses [54]. Diminished levels of selenium in the brain are associated with cognitive decline [55] and Alzheimer's disease [56]. Selenium supplementation has been linked with improvements in mood [57] and protection against postpartum depression [58]. What is unclear is if low dietary selenium is a risk factor for the development of depression. In recognition of selenium's biological activity, it has been hypothesized that low levels of dietary selenium would be associated with an increased risk of major depressive disorder (MDD) in a representative population-based sample of women.

Alterations in redox biology are established in depression; however, there are no prospective epidemiological data on redox-active selenium in depression. It is known that selenium's function as an antioxidant, and as a constituent of selenoproteins that are important in redox homeostasis, warrants further investigation as a risk factor for depression, and suggest a potentially novel modifiable factor in the primary prevention and management of depression.

2.3. Selenium and allergies

The International Study of Asthma and Allergies in Childhood (ISAAC) found that one in four New Zealand children aged 6–7 years had experienced asthma symptoms, which placed New Zealand in the top four countries for asthma prevalence [59]. The reasons for the high prevalence and severity of this condition or the increased prevalence of asthma over the last 20 years are not well understood. One of a number of environmental factors that have been proposed as a reason for the escalation in asthma prevalence is a decreasing intake of dietary antioxidants [60]. It is well known that selenium is essential for the optimal functioning of the selenoenzymes glutathione peroxidases (GPx) and thioredoxin reductases, powerful antioxidants, and is found abundantly in lung tissue and the extracellular fluid of the respiratory system [61]. Selenium has been implicated in inflammation by reducing the severity of the inflammatory response through modulation of the pro-inflammatory leukotrienes, important mediators of acute asthmatic reactions as well as sustaining the inflammatory process causing a late allergic reaction metabolism [62]. Evidence from randomized controlled trials [63] and basic mechanistic work investigating the effect of selenium on markers of inflammation and oxidative stress [62]. Evidences have supported a protective role for selenium in asthma, although other studies have not [64-66]. The ISAAC study does not support a strong association between selenium status and the high incidence of asthma in New Zealand. However, there was a modest association between lower plasma selenium and whole blood glutathione peroxidase activity and higher incidence of persistent wheeze [67].

2.4. Selenium in preventing oxidative stress

The reactivity of organoselenium compounds [22,68] characterized by high nucleophilicity and antioxidant potential, and provides the basis for their pharmacological activities in mammalian models. Organochalcogens have been widely studied given their antioxidant activity, which confers neuroprotection, antiulcer, and antidiabetic properties. Given the complexity of mammalian models, understanding the cellular and molecular effects of organochalcogens has been hampered. In reference [69] the nematode worm *Caenorhabditis elegans* is an alternative experimental model that affords easy genetic manipulations, green fluorescent protein tagging, and in vivo live analysis of toxicity. Manganese (Mn)-exposed worms exhibit oxidative-stress-induced neurodegeneration and life-span reduction. Diethyl-2-phenyl-2-tellurophenyl vinyl phosphonate (DPTVP) and 2-Phenyl-1,2-benzoisoselenazol-3-(2H)-one (Ebselen) were tested for reversing the Mn-induced reduction in survival and lifespan in this nematode. DPTVP was the most efficacious compound as compared to Ebselen in reversing the Mn-induced toxicity and increasing in survival and life span. DPTVP and ebselen act as antiaging agents in a model of Mn-induced toxicity and aging by regulating DAF-16/FOXO signaling and attenuating oxidative stress.

Bone is a specialized connective tissue, which forms the framework of the body. Various physiological conditions can adversely affect femoral bone metabolism. These physiological conditions could be food deprivation [70], and iodine and/or selenium (Se) deficiency [71,72] and antithyroid drugs [73] affects bone maturation. Selenium is an important protective element that may be used as a dietary supplement protecting against oxidative stress, cellular damage and bone impairments [74].

2.5. Selenium in HIV infection

The HIV pandemic has placed a great demand upon the scientific community to develop effective prevention and treatment methods. Since the beginning of the pandemic in 1981, over 25 million people are estimated to have died from the disease [75]. It is currently a leading cause of death in many parts of the world, and a disease that disproportionately affects the marginalized and socially disadvantaged. It is currently a leading cause of death in many parts of the world, and a disease that disproportionately affects the marginalized and socially disadvantaged. Many of those affected also suffer from chronic food insecurity and malnutrition, so therapies that could potentially target both HIV disease and malnutrition, such as multivitamins, have been extensively researched for potential benefits [76]. Among such therapies, the antioxidant micronutrients theorized to have potential benefits in HIV disease, apart from correcting deficiencies, have been examined frequently [77,78].

Selenium has an inhibitory effect on HIV in vitro through antioxidant effects of glutathione peroxidase and other selenoproteins. Numerous studies have reported low selenium status in HIV-infected individuals, and serum selenium concentration declines with disease progression. Some cohort studies have shown an association between selenium deficiency and progression to AIDS or mortality. In several randomized controlled trials, selenium supplementation has reduced hospitalizations and diarrheal morbidity, and improved CD4+ cell counts, but the evidence remains mixed. Additional trials are recommended to study the effect of selenium supplementation on opportunistic infections, and other HIV disease-related comorbidities in the context of highly active antiretroviral therapy in both developing and developed countries [79].

There is a historical record showing that organoselenium compounds can be used as antiviral and antibacterial agents. This topic has been reviewed by [22,23].

2.6. Selenium in the brain

In addition to the well-documented functions of Se as an antioxidant and in the regulation of the thyroid and immune function [80]. Recent advances have indicated a role of Se in the maintenance of brain function [81]. Selenium is widely distributed throughout the body, but is particularly well maintained in the brain, even upon prolonged dietary Se deficiency [82]. In the brain, the highest concentration of Se is found in the gray matter, an area responsible for chemical synaptic communication [83]. It has been shown that rats on a Se-deficient diet for thirteen weeks retained Se in their brain, while their plasma Se concentrations were depleted [84]. After intraperitoneal injection of $^{75}SeO3^{2-}$ into Se-deficient rats, the brain rapidly sequesters a large portion of the available Se [85]. In the brain, it was found that the cerebellum accumulated the highest concentration of Se, followed by the cortex, medulla oblongata, cerebral hemisphere, and the spinal cord. Interestingly, Se retention in the brain depends on Selenoprotein P expression [86]. Because the body preferentially allocates available Se to the brain during Se deficiency, Se may play an essential role in the brain. More evidence for the brain being at the apex of Se retention is provided by a study showing that a six generation Se deficiency in rats caused a more than 99% reduction of Se concentration in the liver, blood, skeletal tissue, and muscle, while the brain retained a 60% of the Se [87]. Se concentration in Alzheimer's brains was found to be 60% of the age-matched control individuals [88]. Accumulated lines of evidence indicate important roles of selenoproteins in the maintenance of optimal brain functions via redox regulation. Decreased expression of several selenoproteins is associated with the pathologies of a few age-associated neurodisorders, including Parkinson's disease, Alzheimer's disease and epilepsy [81].

Oxidative stress and generation of reactive oxygen species are strongly implicated in a number of neuronal and neuromuscular disorders, including epilepsy. The functions of selenium as an antioxidant trace element are believed to be carried out by selenoproteins that possess antioxidant activities and the ability to promote neuronal cell survival [89]. It is known the role of selenium in a detoxifying enzyme, glutathione peroxidase, this element has been demonstrated to have a positive biological function in various aspects of human health [90]. Oxidative stress and generation of reactive oxygen species are strongly implicated in a number of neurologic disorders including seizure disorders. Oxidative phosphorylation occurring in the mitochondria produces oxygen radicals routinely in all tissues as well as the nervous system. One important defense may be to remove the oxygen radicals. Selenium-requiring processes are involved in normal maintenance of cell function. However, when the system is overused or chronically activated beyond its normal state, such as recurrent or intractable seizures, abnormal increases in by-products can produce neuronal cell damage. Selenium provides protection from reactive oxygen species–induced cell damage. The proposed mechanisms are mainly through the functions of seleno-dependent enzymes and selenoproteins [82,91]. It seems that selenium plays an important role in stopping the vicious cycle of oxidative stress and neuronal damage in patients with intractable seizures by restoring the defense mechanism.

2.7. Selenium and the thyroid

Some selenoproteins of the human selenoproteome display multiple genes performing similar functions. The main selenoprotein families are the glutathione peroxidases (GPxs; seven genes), the thioredoxin reductases (TRxs; three genes) and the iodothyronine deiodinases (DIs; three genes) [92,93]. The GPxs, which possess oxidoreductase functions, protect the cell from oxidative stress. The TRxs form a cellular redox system, existing in many organisms, which is essential for cell development and proliferation. The DIs that catalyzes the conversion of T4 to T3 provides the sources of T3 production. It may thus be hypothesized that the essential micronutrient selenium, in the form of Se-Cys, modulates redox-sensitive signaling pathways and thereby potentially modifies selenoprotein gene expression. These findings have aroused growing interest of the scientific community in this multifaceted element. In this context, whereas selenium administration for cancer chemoprevention produced questionable results, those of selenium supplementation in patients with autoimmune thyroid disease have been more encouraging. In [94] comprises an in-depth discussion of the link between selenium and thyroid function; it provides a critical analysis of the data contained in recent studies, an update and evaluation of current knowledge with regard to the mechanisms of action of selenium, and reflections on the prospects for selenium supplementation in thyroid pathology.

Evidence in support of selenium supplementation in thyroid autoimmune disease is evaluated; the results herein presented demonstrating the potential effectiveness of selenium in reducing the antithyroid peroxidase titer and improving the echostructure in the ultrasound examination. However, considerable discord remains as to who should comprise target groups for selenium treatment, who will most benefit from such treatment, the precise impact of the basal antithyroid peroxidase level, and the effect of disease duration on the treatment outcome. Clearly, further in-depth studies and evaluation are required concerning the mechanism of action of selenium as well as the choice of supplements or dietary intake.

2.8. Selenium in cancer

The reactive oxygen species (ROS) are derived from cellular oxygen metabolism and from exogenous sources. An excess of ROS results in oxidative stress and may eventually cause cell death. ROS levels within cells and in extracellular body fluids are controlled by concerted action of enzymatic and non-enzymatic antioxidants. The essential trace element selenium exerts its antioxidant function mainly in the form of selenocysteine residues as an integral constituent of ROS-detoxifying selenoenzymes such as glutathione peroxidases (GPx), thioredoxin reductases (TrxR) and possibly selenoprotein P (SeP). In particular, the dual role of selenoprotein P as selenium transporter and antioxidant enzyme is highlighted herein. A cytoprotective effect of selenium supplementation has been demonstrated for various cell types including neurons and astrocytes as well as endothelial cells. Maintenance of full GPx and TrxR activity by adequate dietary selenium supply has been proposed to be useful for the prevention of several cardiovascular and neurological disorders. On the other hand, selenium supplementation at supranutritional levels has been utilized for cancer prevention: antioxidant selenoenzymes as well as prooxidant effects of selenocompounds on tumor cells are thought to be involved in the anti-carcinogenic action of selenium [95,96].

Among various antioxidant minerals, selenium it may prove to be of major significance as a prophylactic agent against cancer. Low blood selenium concentration and incidence of carcinogenesis have been well observed in both animals [97] as well as in human studies [98]. In addition, it has been demonstrated in a double blind randomized cancer prevention trial in humans that increased selenium intake has a significant role in the treatment of cancer [99]. A similar prospective study could also be designed for other cancers to determine the chemopreventive effect of Se. Selenium has also been reported to have a beneficial effect on the incidence of gastrointestinal and bladder cancers [100,101].

Although selenium is reported to play a significant role in cancer development, its exact anticancer mechanism of action at molecular levels is not fully understood. However, it has been hypothesized that the most possible mechanistic action of Se as chemoprevention is its role in the antioxidant defense systems to reduce oxidative stress and limit DNA damage [24,102]. Experiments carried out within the framework of a canine model using male beagle dogs to mimic prostate cancer in humans showed that the damage to DNA was significantly reduced when the animals were exposed to increased Se dietary supplements [103]. The effectiveness of Se in the prevention of DNA damage, however, depends on its chemical forms. In an *in vitro* study [104] found that selenocysteine inhibited DNA damage more strongly than the selenomethionine. Other possible anticancer mechanisms of Se include the induction of apoptosis, cell-cycle arrest and DNA-repair genes, inhibition of protein kinase C activity and cell growth and effect on estrogen- and androgen-receptor expression [102].

In [105] knowledge of the plasma selenium levels are associated with optimized concentration or activity of specific selenoproteins can provide considerable insights from epidemiological data on the possible involvement of those selenoproteins in health, most notably with respect to cancer. For cohort studies, if selenoproteins such as glutathione peroxidase and selenoprotein P are relevant to cancer, one might only expect to see an effect on risk when the concentrations in the cohort range from below, to above, the level needed to optimize the activity or concentration of these enzymes. Similarly, trials would only show a beneficial effect of supplementation if selenium status were raised from below, to above, the optimal concentration for the selenoproteins likely to be implicated in cancer risk, as occurred in the Nutritional Prevention of Cancer (NPC) trial but not in Selenium and Vitamin E Cancer Prevention Trial (SELECT). The most powerful evidence for the involvement of selenoproteins in human health comes from epidemiological studies that have related single nucleotide polymorphisms in selenoproteins to disease risk. The totality of the evidence currently implicates GPx1, GPx4, SEPS1, Sep15, SEPP1 and TXNRD1 in conditions such as cardiovascular disease, pre-eclampsia and cancer. Future studies therefore need to determine not only selenium status, but genotype, both in selenoproteins and related pathways, when investigating the relationship of selenium with disease risk.

2.9. Selenium in diabetes

The evidence supporting an effect of selenium on the risk of diabetes is variable, occasionally conflicting, and limited to very few human studies. Following a trial investigating the effect of selenium supplementation (200 μg/day) on skin cancer, subsequent analysis showed that there was an increased risk of developing type 2 diabetes in the supplemented group. Evidence from analysis of NHANES III [106] supports these findings; the adjusted mean serum selenium concentrations were slightly, but significantly, higher in diabetics compared with those without the disease. This study, conducted in an elderly French population, found a sex-specific protective effect of higher selenium status at baseline on later occurrence of dysglycemia; that is, risk of dysglycemia was significantly lower in men with plasma selenium, but no significant relationship was observed in women [107].

The role of selenium as an antioxidant, particularly within the GPxs, selenium is likely to be important in reducing oxidative stress, an important risk factor for developing diabetes. There are also plausible suggestions that selenium can influence glucose metabolism. However, at high intakes it is also conceivable that reactive oxygen species could be generated or selenium may accumulate in the organs associated with glucose metabolism [108]. In patients with diabetes, selenium supplementation (960 μg/day) reduced NF-κB levels to those comparable with nondiabetic controls [109]. In addition, further analysis of the Nutritional Prevention of Cancer trial data has shown an increased risk of self-reported Type-2 diabetes in those supplemented with Se, though the effect was significant only in those in the top tertile of plasma Se at baseline [110].

2.10. Selenium and male fertility

Selenoprotein P transports selenium particularly to testis and brain [111]. Among the five enzymes of GPx, GPx1 prevents apoptosis induced by oxidative stress and GPx4 acts directly on membrane phospholipid hydroperoxides and detoxifies them. Selenium as GPx, is present in spermatids and forms the structural part in the mid piece of mature spermatozoa. Some well known effects of selenium deficiency include instability of the middle piece leading to defective sperm motility [112], low reproductive ability and abnormal development of spermatozoa [113]. Selenium is also required for testosterone synthesis and sequential development of flagella [114]. It can restore the physiological constitution of polyunsaturated fatty acid in the cell membrane [115]. Testes are extremely resistant to Se depletion and have high Se content. Recent studies have shown that sperm and testicular Se was unaffected by the supplementation, suggesting that testes are protected from Se excess as well as from Se deficiency [116].

2.11. Selenium in asthma

Se status is decreased in patients with asthma, as is activity of glutathione peroxidase in platelets and erythrocytes. There is an associated marked oxidant/antioxidant imbalance in the blood of asthmatics, which reflects poor antioxidant status and enhanced inflammatory mediated oxidative stress [117]. According to the University of Maryland Medical Center, a 2004 study of 24 asthmatics that were given selenium supplements for 14 weeks had significant improvement in their symptoms when compared to a control group given a placebo. Although this is a small study done over a short amount of time, it's encouraging [118].

2.12. Selenium in cardiovascular disorders

Free radicals are toxic to the myocardium and can cause tissue damage that leads to extensive necrosis, myocytolysis and cellular edema [119]. Atherosclerotic plaque formation may be a reflection of sub-optimal GPx4 activity in the prevention of LDL oxidation, with subsequent uptake by endothelial cells and macrophages in arterial blood vessels [120]. Selenium via GPx reduces phospholipids, hydro peroxides and cholestryl esters associated with lipoproteins and may therefore, not only reduce the accumulation of oxidized LDL in arterial wall but also reduce platelet aggregation and activation of monocyte and macrophages [121]. Selenium owing to its antithrombotic effect on the interaction between platelets and endothelial cells via GPx, also provides concrete evidence in the prevention of atherosclerosis [122].

The study on acute myocardial infarction (AMI) patients, it was observed that selenium dependent GPx level decreases significantly in AMI patients and explained it as an imperative consequent of GPx activity in annihilating oxygen toxicity by metabolizing H_2O_2 and inhibiting further free oxygen radical production in early phase of myocardial infarction [123].

2.13. Selenium in rheumatoid arthritis

Scientific research shows that people with rheumatoid arthritis have low levels of selenium. A study suggests, it is part of the body's defense mechanism [124]. In reference [125] the authors found lower selenium levels in patients with rheumatoid arthritis who were treated with arthritis medication compared with people without the condition. In people without rheumatoid arthritis or a family history of the condition, low levels of the mineral may increase the risk of developing rheumatoid arthritis [126].

2.14. Selenium in pre-eclampsia

In reference [124], pre-eclampsia (pregnancy induced hypertension; PIH), is an important cause of maternal morbidity and mortality with essentially unknown etiology. However, the precise factors involved in the pathogenesis of PIH are still unknown [127]. It has been conceived that free radical mediated oxidative stress may contribute to the development of pre-eclampsia. Selenium and its related enzymes especially GPx play a crucial role in annihilating oxygen toxicity and there by controlling the progression of disease [128]. In addition, selenium deficiency in women may result in infertility, miscarriages and retention of the placenta [129].

2.15. Selenium in immunity

In reference [130], the generation of ROS in a limited dose is one of the processes induced by the immune system to destroy microbial pathogens and viruses. However, the over-production of ROS can also cause damage to the host cells that need to be protected by Se at various stages in the immune system. Keshan disease, an endemic cardiomyopathy in China that develops as a result of Se deficiency, may also be complicated with viral infection, and this has led to the investigation of the effects of viruses, such as coxsackievirus, on Se-deficient animals [131,132]. Results from animal studies have demonstrated that Se deficiency can lead to an impairment of immune functions that result in the inability of phagocytic neutrophils and macrophages to destroy antigens. A low Se status in humans has been reported to cause a decreased immune response to poliovirus vaccination [133]. This study also demonstrated that the subjects supplemented with Se showed fewer mutations in poliovirus than those who received a placebo. The involvement of Se in the immune system may be associated with a number of mechanisms, including the increased activity of natural killer (NK) cells, the proliferation of T-lymphocytes, increased production of interferon c, increased high-affinity interleukin-2 receptors, stimulation of vaccine-induced immunity and increased antibody-producing B-cell numbers [134,135].

2.16. Selenium in bone impairments

Osteoblasts (bone-forming cells) and osteoclasts (bone-resorption cells) are involved in bone remodeling. Therefore, any loss of osteoblastic activity or an increase in osteoclastic activity could lead to a decrease in bone-mineral densities (BMD), bone mass, and make the bones more likely to osteoporosis, and ultimately to fractures [136]. In addition, high levels of reactive oxygen species (ROS) and many other factors such as genetic race, hormonal, mechanical, and nutritional statues are involved in bone weakness and fractures. ROS shift cells into a state of oxidative stress [137] which contributes to the etiology of various degenerative diseases that cause tissue injury [136,137]. Studies have demonstrated that the ischemia-reperfusion processes that occur after a fracture are associated with oxidative stress development [136,138]. It is believed that bone markers such as osteocalcin and alkaline phosphatase as well as antioxidant enzymes play a significant role in fracture healing. However, to the best of our knowledge, there are no reports about the use of vitamins A, C, E, and selenium as antioxidant therapy to explore their effects in the levels of bone-healing markers and oxidative stress parameters of osteoporotic patients. In [139] suggests that selenium is an important protective element that may be used as a dietary supplement protecting against bone impairments.

3. Vitamin E and Health

3.1. Vitamin E

All forms of vitamin E meet the chemical definition of an antioxidant moiety: "chain-breaking free radical scavenger." Indeed consistent data have shown that all isoforms act as potent antioxidants in conventional in vitro paradigms. The free hydroxyl group on the aromatic ring is thought to be responsible for this property, and a relatively stable form of the original vitamin E is formed when hydrogen from this group is donated to a free radical. Yet, definitive proof that vitamin E possesses antioxidant properties has been hampered for a long time because of a lack of sensitive and specific analytical techniques to measure this biologic event *in vivo* [36]. Apart from antioxidant properties, more recent studies have clearly demonstrated that vitamin E also possesses important non-antioxidant cellular and molecular functions. One of the first roles of α-tocopherol in cell signaling was the report that it inhibits smooth muscle cell proliferation, decrease protein kinase C activity, and controls expression of the α-tropomyosin gene [140].

One of the major vitamin E-deficiency symptoms are neurological disorders. Furthermore, vitamin E deficiency is related to female infertility. The frame to pinpoint the physiological action of vitamin E is set by its chemical nature: (i) It is a redox-active compound prone to undergo 1- and 2-electron transitions and (ii) it is highly lipophilic, although this property may be modulated by phosphorylation [141]. In [142] oxidative stress is a developing research field and is being examined in female infertility. Prooxidants, also called free radicals or reactive oxygen species (ROS), and their neutralizing agents the antioxidants are the main chemicals of the oxidation mechanism. The term oxidative stress refers to the dysequilibrium between the free radicals and the antioxidants in favor of the free radicals. In actuality, free radicals are not so frightening, since they are necessary for the adequate reproductive functions within the ovary and the endometrium. Vit E administration may improve the endometrial response in unexplained infertile women *via* the likely antioxidant and the anticoagulant effects. It may also modulate the antiestrogenic effect of clomiphene citrate and the problem of a thin endometrium in these cycles may be adjusted.

Some non-antioxidant properties of vitamin E could play a key role in neuroprotection. It has been recently shown that α-tocotrienol, at nanomolar concentrations, protects mouse hippocampal and cortical neurons from cell death by modulating neurodegenerative signaling cascades. Furthermore, it has been shown that α-tocotrienol modulates 12-lipoxygenase and phospholipase A2 activities, which are implicated in glutamate-induced neuronal cell death [143]. Some vitamin E forms (α- and γ-tocopherol, tocotrienols) also exhibit potent anti-inflammatory properties [144,145]. The introduction of the free radical theory of brain aging has propelled a renewed interest in this vitamin. As result, by preventing and/or minimizing the oxidative stress dependent brain damage, this vitamin plays important role in brain aging, cognition, and Alzheimer's dementia.

Vitamin E is a potent peroxyl radical scavenger that prevents lipid peroxidation [146] and is found in high concentrations in immune cells [147]. Deficiency in vitamin E is associated with increased oxidative stress [148] and impaired immune function, including both humoral and cell-mediated immunity, phagocyte function, and lymphocyte proliferation [149]. Age-related declines in immune function can be restored by vitamin E supplementation [150]. This vitamin is an exogenous, lipidsoluble antioxidant molecule. It is thought to be a direct free radical scavenger by activating the intracellular antioxidant enzymes and saving the cell membranes from lipid peroxidation, which was demonstrated on sperm membrane components [151]. Its antioxidant effect was concluded in cancer therapy, high-risk pregnancy and male infertility [152-154].

Vitamin E (α–tocopherol acetate) is found within the phospholipid bilayer of cell membranes where it functions as an electron donor to free radicals. It has been recognized as one of the body's major natural antioxidants. Another antioxidant, Se appears to function as an antimutagenic agent, preventing the malignant transformation of normal cells. Its protective effects seem to be primarily associated with its presence in the seleno-enzymes which are known to protect DNA and other cellular components from oxidative damage [44].

Selenium, vitamin A (retinol) and vitamin E (α-tocopherol) are essential micronutrients for human health. Both selenium and vitamin E are important in host antioxidant defense and immune function. It has been reported that deficiency of selenium and vitamins may promote peroxidation events leading to the release of free radicals. All have free-radical-scavenging properties that allow them to function as physiologic antioxidants in protecting a number of chronic diseases, such as cancer and cardiovascular disease. In addition to its antioxidant capacity, α-tocopherol regulates expression of genes involved in a wide range of cell functions, including cell cycle regulation, inflammation and cell adhesion, cell signaling, and lipid uptake [155].

Selenium also has an important role in antioxidant defense and immune function. Due to its incorporation as selenocysteine into glutathione peroxidase (GPX) [156] and thioredoxin reductase [157], selenium is important for the control of oxidative stress and, therefore, the redox tone of the cell. In total, there are 25 identified selenoproteins (24 in rodents), many with unknown function [25]. Selenium is important for cytotoxic T-lymphocyte and natural killer cell activity [158], respiratory burst [159], and protection against endotoxin-induced oxidative stress [160]. Multiple studies have shown that NF-kB activation can be affected by selenium status [161,162], and selenium deficiency can alter chemokine and cytokine expression during viral infections [163]. Various investigators have reported the role of selenium as an inhibitor of carcinogenesis in various organs including liver, skin, stomach, mammary gland, gastrointestinal and oral cavity [164,165].

In [166] blader cancer represents an important cause of morbidity and mortality. In 2010 it was again the second most common genitourinary cancer in the United States with an established 70,530 new cases and 14,680 deaths [167]. Currently it is estimated that more than 500,000 men and women in the United States have a history of bladder cancer. The etiology of most bladder urothelial carcinoma is associated with tobacco exposure, occupational exposure to aromatic amines, and exposure to the chemical and rubber industries [168]. Bladder cancer is the most expensive cancer in the United States, accounting for almost $3.7 billion (2001 value) in direct costs [169]. There is substantial epidemiological and biological evidence that selenium and vitamin E may prevent bladder cancer. A recent meta-analysis of 7 published epidemiological studies, including 3 case-control, 3 nested case-control and 1 case cohort series, examined the association between selenium levels and bladder cancer [170]. In the analysis stratified by gender only women showed a significantly decreased risk associated with selenium. An opposite gender pattern, with protective effects in men but not in women, was reported in a meta-analysis of selenium supplementation, primary cancer incidence and mortality [171]. Epidemiological and biological evidence suggests a preventive effect of selenium and vitamin E on bladder cancer. These researches assessed the effect of selenium and/or vitamin E on bladder cancer development.

3.2. Selenium and vitamin E

Selenium and vitamin E are essential components of the human diet and have been studied as antioxidants and/or potential agents for a variety of human diseases. Various formulations of both selenium and vitamin E have been shown to possess a therapeutic and preventive effect against prostate cancer. The Selenium and Vitamin E Cancer prevention Trial (SELECT) started in 2001 and was a phase III, randomized placebo/controlled human trial to investigate the prostate cancer chemopreventive effects of selenium and vitamin E or their combination [172,173].

Sselenium an essential trace element, and vitamin E, a lipid soluble antioxidant, are important mediators for protection against oxidative stress. Deficiencies in either Se or vitamin E result in increased viral pathogenicity and altered immune responses. Furthermore, deficiencies in either Se or vitamin E results in specific viral mutations, changing relatively benign viruses into virulent ones. Thus, host nutritional status should be considered a driving force for the emergence of new viral strains or newly pathogenic strains of known viruses [174].

Several studies have evaluated the possible association between antioxidants vitamins or selenium supplement and the risk of prostate cancer, but the evidence is still inconsistent. We systematically searched PubMed, EMBASE, the Cochrane Library, Science Citation Index Expanded, Chinese biomedicine literature database, and bibliographies of retrieved articles up to January 2009. We included 9 randomized controlled trials with 165,056 participants; methodological quality of included trials was generally high. Meta-analysis showed that no significant effects of supplementation with β-carotene (3 trials), vitamin C (2 trials), vitamin E (5 trials), and selenium (2 trials)versus placebo on prostate cancer incidence. The mortality of prostate cancer did not differ significantly by supplement of β-carotene (1 trial), vitamin

C (1 trial), vitamin E (2 trials), and selenium (1 trial). This study indicates that antioxidant vitamins and selenium supplement did not reduce the incidence and mortality of prostate cancer; these data provide no support for the use of these supplements for the prevention of prostate cancer [175].

Epidemiological studies demonstrated that human exposure to methylmercury (MeHg) may contribute to the development and progression of metabolic and cardiovascular disorders. However, the mechanisms involved and the role of selenium (Se) and vitamin E (VE) supplementation in modulating MeHg cardiovascular toxicities remain unclear. The effects of Se and VE supplementation on MeHg-mediated systemic oxidative stress, antioxidant defense, inflammation, and endothelial dysfunction are carried out in an animal model. Male Sprague–Dawley rats were fed a starch-based casein diet or the same diet supplemented with 1 or 3 mg Se/kg diet and with or without 250 or 750 mg VE/kg diet. After 28 days of dietary treatment, rats were gavaged with 0 or 3 mg MeHg/kg BW for 14 consecutive days. Results suggested that exposure to MeHg may increase the risk of cardiovascular disease by decreasing circulating paraoxonase-1 activities, increasing serum oxidized low density lipoprotein levels, and associated systemic inflammation and endothelial dysfunction as reflected by increased leukocyte counts and serum levels of intercellular adhesion molecule-1 and monocyte chemotactic protein-1. Se and VE supplementation may either alleviate or augment the effects of MeHg, depending on their doses and combinations [176].

The analysis of the hepatotoxic effect of malathion in adult male rats and evaluate the possible hepatoprotective effect of vitamin E and/or selenium. Oral administration of malathion for 45 days significantly induced marked hepatic injury as revealed by increased activity of the plasma enzymes (alanine aminotransferase (ALT), aspartate aminotransferase (AST), lactate dehydrogenase (LDH) and gamma-glutamyl transferase GGT). Oral administration of vitamin E and selenium in combination with malathion exhibited a significant protective effect by lowering the elevated plasma levels of the previous enzymes. Light microscopic investigation revealed that malathion exposure was associated with necrosis of hepatocytes, marked changes of liver tissues in the form of dilated veins, hemorrhagic spots and some degenerative signs of hepatocytes [177].

4. Conclusion

Research on Se during the last few years has produced a great deal of evidence demonstrating the important role that Se and its metabolites play in human diseases. In particular, our knowledge of the functional roles of the GPx and TrxR groups as essential antioxidant selenoenzymes in protecting cells from oxidative stress has greatly increased, as has the link between these enzymes and various diseases. However, there are still areas of research that require in-depth study, including the mechanistic modes of action of Se in cancer etiology, how Se delivers its anticancer activity at the molecular and genetic levels, and what biomarkers can be used to accurately measure the efficacy of Se for use in chemoprevention. It is not well understood the specific mechanism by which Se protects cells and tissue at the cellular level from damage due to oxidative stress; this is particularly relevant in heart diseases, which are still a major cause of death worldwide. Given the number of Se cancer preventive trials that are currently being undertaken in many countries, the significant outcomes of these trials will not only provide us with more information on optimal Se intake for the treatment and prevention of cancer, but they will also provide us with strategies in the management of other potential human diseases associated with low Se status. Until the specific biomarkers are identified that will directly link Se with disease prevention and treatment, its use as supplements in health therapy should be taken with caution.

The Selenium and Vitamin E Cancer prevention Trial (SELECT) failed to show an effect in human population. However, pre- and pro-SELECT studies are still supporting the potential usefulness of selenium and/or vitamin E for prevention of prostate cancer and possibly other conditions. Much remains to be understood about the absorption, metabolism and physiologic chemistry of these agents. Nonetheless, the existing evidence supporting selenium and vitamin E as potential prostate cancer chemopreventive agents is possibly enough to justify further efforts in this direction.

My goal in putting this review together was to provide a wide range of subjects dealing with selenium and vitamin E supplementation, that are used in chronic disease prevention, due to their antiradical activities indicating that the combine effects of Se and vitamin E could provide an important dietary source of antioxidants and/or potential agents for a variety of human diseases. It is my hope that readers will find this chapter to be useful in further studies dealing with this subject.

Author details

Manuel Soriano García*

Chemistry of Biomacromolecules Department, Chemistry Institute, National Autonomus University of México, University City, Mexico

References

[1] Berzelius, J. J. (1818). Afhandl. Fys. Kemi Minerag. 6, 42

[2] Shamberger, R. J. (1984). Selenium. *In Biochemistry of the Essential Ultratrace Elements*, Earl Frieden, Ed.; Plenum Press; New York, NY, 3, 201-237.

[3] Brasted, R. C. (1961). *In Comprehensive Inorganic Chemistry: Sulfur. Selenium, Tellurium, Polonium and Oxygen*, Robert C. Brasted, Ed.; D. Van Nostrand Co.; Princeton, NJ, 2.

[4] Merian, E. (1984). Introduction on environmental chemistry and global cycles of chromium, nickel, cobalt beryllium, arsenic, cadmium and selenium, and their derivatives. *Toxicological and Environmental Chemistry*, 8, 9-38.

[5]   Smith, M., Franke, K. W., & Westfall, B. B. (1936). The selenium problem in relation
      to public health. A preliminary survey to determine the possibility of selenium intox-
      ication in the rural population living in seleniferous soil. *US Public Health Reports*, 51,
      1496-1505.

[6]   Thomson, C. D. (2004). Assessment of requirements for selenium and adequacy of se-
      lenium status: a review. *European Journal of Clinical Nutrition*, 58, 391-402.

[7]   Goldhaber, S. B. (2003). Trace element risk assessment vs toxicity. *Regulatory Toxicolo-
      gy and Pharmacology*, 38, 232-242.

[8]   Castellano, S., Gladyshev, V. N., Guigo, R., & Berry, M. J. (2008). SelenoDB 1.0: a da-
      tabase of selenoprotein genes, proteins and SECIS elements. *Nucleic Acids Research*,
      36, 332D-338D.

[9]   Allmang, C., Wurth, L., & Krol, A. (2009). The selenium to selenoprotein pathway in
      eukaryotes: more molecular partners than anticipated. *Biochemica et Biophysica Acta*,
      1790, 1415-1423.

[10] Arbogast, S., & Ferreiro, A. (2009). Selenoproteins and protection against oxidative
      stress selenoprotein N as a novel player at the crossroads of redox signaling and cal-
      cium homeostasis. *Antioxidants & Redox Signaling*, 12, 893-904.

[11] Kipp, A., Banning, A., Van Schothorst, E. M., Meplan, C., Schomburg, L., Evelo, C.,
      Coort, S., Gaj, S., Keijer, J., Hesketh, J., & Brigelius-Flohe, R. (2009). Four selenopro-
      teins, protein biosynthesis, and Wnt signaling are particularly sensitive to limited se-
      lenium intake in mouse colon. *Molecular Nutrition & Food Research*, 53, 1561-1572.

[12] Lescure, A., Rederstorff, M., Krol, A., Guicheney, P., & Allamand, V. (2009). Seleno-
      protein function and muscle disease. *Biochimica et Biophysica Acta*, 1790, 1569-1574.

[13] Lobinski, R., Edmonds, J. S., Sukuki, K. T., & Uden, P. C. (2000). Species-selective de-
      termination of selenium compounds in biological materials. *Pure and Applied Chemis-
      try*, 72, 447-461.

[14] Bleys, J., Navas-Acien, A., Stranges, S., Menke, A., Miller, E. R., III, & Guallar, E.
      (2008). Serum selenium and serum lipids in US adults. *The American Journal of Clinical
      Nutrition*, 88, 416-423.

[15] Johnson, C. C., Fordyce, F. M., & Rayman, M. P. (2010). Symposium on "geographical
      and geological influences on nutrition": factors controlling the distribution of seleni-
      um in the environment and their impact on health and nutrition. *Proceedings of the
      Nutrition Society*, 69, 119-132.

[16] Navarro-Alarcon, M., & Cabrera-Vique, C. (2008). Selenium in food and the human
      body: a review. *Science of the Total Environment*, 400, 115-141.

[17] Schrauzer, G. N., & Surai, P. F. (2009). Selenium in human and animal nutrition: re-
      solved and unresolved issues. A partial historical treatise in commemoration of the
      fiftieth anniversary of the discovery of the biologically essentiality of selenium, dedi-

cated to the memory of Klaus Schwarz (1914-1978) on the occasion of the thirtieth anniversary of his dead. *Critical Reviews in Biotechnology*, 29, 2-9.

[18] World Health Organization. (1987). Selenium. *Geneva; WHO*.

[19] Thompson, H. J. (2001). *In Selenium: Its Molecular Biology and Role in Human Health*, Dolph L. Hatfield, Ed.; Kluwer Academic. Publishers; Boston, MA, 283-297.

[20] Hatfield, D. L. (2001). *In Selenium: Its Molecular Biology and Role in Human Health*, Dolph L. Hatfield, Ed.; Kluwer Academic. Publishers; Boston, MA.

[21] Bols, M., Lopez, O., & Ortega-Caballero, F. (2007). In Kamerling, J. P., Ed., *Comprehensive Glycoscience: From Chemistry to Systems Biology*, Elsevier Science: Oxford, 1, 815-884.

[22] Soriano-Garcia, M. (2004). Organoselenium Compounds as Potential Therapeutic and Chemopreventive Agents: A review. *Current. Medicinal Chem*, 11, 1657-1169.

[23] Mugesh, G., du Mont, W.-W., & Sies, H. (2001). Chemistry of Biologically Important Synthetic Organoselenium Compounds. *Chemical Reviews*, 101, 2125-2180.

[24] Rayman, M. P. (2005). Selenium in cancer prevention: a review of the evidence and mechanism of action. *Proceedings of the Nutrition Society*, 64, 527-542.

[25] Papp, L. V., Holmgren, A., & Khanna, K. K. (2010). Selenium and Selenoproteins in Health and Disease. *Antioxidants & Redox Signaling*, 12, 793-795.

[26] Jablonska, E., Gromadzinska, J., Sobala, W., Reszka, E., & Wasowicz, W. (2008). Lung cancer risk associated with selenium status is modified in smoking individuals by Sep15 polymorphism. *European Journal of Nutrition*, 47, 47-54.

[27] Kryukov, G. V., Castellano, S., Novoselov, S. V., Lobanov, A. V., Zehtab, O., Guigo, R., et al. (2003). Characterization of mammalian selenoproteomes. *Science*, 300, 1439-1443.

[28] Ashton, K., Hooper, L., Harvey, L. J., Hurst, R., Casgrain, A., & Fairweather-Tait, S. J. (2009). Methods of assessment of selenium status in humans: a systematic review. *The American Journal of Clinical Nutrition*, 89, 2025S-2039S.

[29] Ip, C., Dong, Y., & Ganther, H. E. (2002). New concepts in selenium chemoprevention. *Cancer and Metastasis Reviews*, 21, 281-289.

[30] Schrauzer, G. N. (2009). Selenium and selenium-antagonistic elements in nutritional cancer prevention. *Critical Reviews in Biotechnology*, 29, 10-17.

[31] Evans, P., & Halliwell, B. (2001). Micronutrients: oxidant/antioxidant status. *British Journal of Nutrition*, 85, 67S-74S.

[32] Wolf, G. (2005). The discovery of the antioxidant function of vitamin E: the contribution of Henry A. Matill. *Journal of Nutrition*, 135, 363-366.

[33] Sies, H., & Murphy, M. E. (1991). Role of tocopherols in the protection of biological systems against oxidative damage. *Journal of Photochemistry and Photobiology*, 8, 211-218.

[34] Brigelius-Flohe', R., & Traber, M. G. (1999). Vitamin E: function and metabolism. *The FASEB Journal*, 13, 1145-1155.

[35] Yap, S. P., Yuen, K. H., & Wong, J. W. (2001). Pharmacokinetics and bioavailability of alpha-, gamma-, and delta-tocotrienols under different food status. *Journal of Pharmacy and Pharmacology*, 53, 67-71.

[36] Yash, B. J., & Pratico, D. (2012). Vitamin E in aging, dementia and Alzheimer's disease. *Biofactors*, 38, 90-97.

[37] Hacquebard, M., & Carpentier, Y. A. (2005). Vitamin E: absorption, plasma transport and cell uptake. *Current Opinion in Clinical Nutrition & Metabolic Care*, 8, 133-138.

[38] O'Byrne, D., Grundy, S., Packer, L., Devaraj, S., Baldenius, K., Hoppe, P. P., Kraemer, K., Jialal, I., & Traber, M. G. (2000). Studies of LDL oxidation following alpha-, gamma-, or delta-tocotrienyl acetate supplementation of hypercholesterolemic humans. *Free Radical Biology & Medicine*, 29, 834-845.

[39] Horiguchi, M., Arita, M., Kaempf-Rotzoll, D. E., Tsujimoto, M., Inoue, K., & Arai, H. p. (2003). pH-dependent translocation of alpha-tocopherol transfer protein (alpha-TTP) between hepatic cytosol and late endosomes. *Genes Cells*, 8, 789-800.

[40] Uchida, T., Abe, C., Nomura, S., Ichikawa, T., & Ikeda, S. (2012). Tissue distribution of alfa- and gamma- tocotrienol and gama-tocopherol in rats and interference with their accumulation by alpha-tocopherol. *Lipids*, 47, 129-139.

[41] Kokarnig, S., Kuehnelt, D., Stiboller, M., Hartleb, U., & Francesconi, K. A. (2011). Quantitative determination of small selenium species in human serum by HPLC/ ICPMS following a protein-removal, pre-concentration procedure. *Analytical & Bioanalytical Chemistry*, 400, 2323-2327.

[42] Suzuki, K. T., Kurasaki, K., Okazaki, N., & Ogra, Y. (2005). Selenosugar, trimethylselenonium among urinary Se metabolites: dose- and agerelated changes. *Toxicology and Applied Pharmacology*, 206, 1-8.

[43] Klotz, L. O., Kroncke, K. D., Buchczyk, D. P., & Sies, H. (2003). Role of copper, zinc, selenium, tellurium in the cellular defense against oxidative and nitrosative stress. *Journal of Nutrition*, 133, 1448S-1451S.

[44] Valko, M., Rhodes, C. J., Moncol, J., Izakovic, M., & Mazur, M. (2006). Free radicals, metals, antioxidants in oxidative stress-induced cancer, *Chemico-Biological Interactions*, 160, 1-40.

[45] Susan, J. F. T., Yongping, B., Martin, R. B., Rachel, C., Dianne, F., John, E. H., et al. (2011). Selenium in human health and disease. *Antioxidants & Redox Signaling*, 14, 1337-1383.

[46] Pasco, J. A., Jacka, F. N., Williams, L. J., Evans-Cleverdon, M., Sharon, L., Brennana, S. L., Kotowicza, M. A., Nicholsone, G. C., Ball, M. J., & Berk, M. (2012). Dietary selenium and major depression: a nested case-control study. *Complementary Therapies in Medicine*, 20, 119-123.

[47] Maes, M., Galecki, P., Chang, Y. S., & Berk, M. (2011). A review on the oxidative and nitrosative stress (O&NS) pathways in major depression and their possible contribution to the (neuro) degenerative processes in that illness. *Progress in Neuropsychopharmacology & Biological Psychiatry*, 35, 676-692.

[48] Ng, F., Berk, M., Dean, O., & Bush, A. I. (2008). Oxidative stress in psychiatric disorders: evidence base and therapeutic implications. *The International Journal of Neuropsycho- pharmacology*, 11, 851-876.

[49] Berk, M., Ng, F., Dean, O., Dodd, S., & Bush, A. I. (2008). Glutathione: a novel treatment target in psychiatry. *Trends in Pharmacological Sciences*, 29, 346-351.

[50] Herken, H., Gurel, A., Selek, S., Armutcu, F., Ozen, M. E., Bulut, M., et al. (2007). Adenosine deaminase, nitric oxide, superoxide dismutase, and xanthine oxidase in patients with major depression: impact of antidepressant treatment. *Archives of Medical Research*, 38, 247-252.

[51] Sarandol, A., Sarandol, E., Eker, S. S., Erdinc, S., Vatansever, E., & Kirli, S. (2007). Major depressive disorder is accompanied with oxidative stress: short-term antidepressant treatment does not alter oxidativeantioxidative systems. *Human Psychopharmacology*, 22, 67-73.

[52] Whanger, P. D. (2001). Selenium and the brain: a review. *Nutritional Neuroscience*, 4, 81-97.

[53] Machado, M. S., Rosa, R. M., Dantas, A. S., Reolon, G. K., Appelt, H. R., Braga, A. L., et al. (2006). An organic selenium compound attenuates apomorphine-induced stereotypy in mice. *Neuroscience Letters*, 410, 198-202.

[54] Malhi, G. S., & Berk, M. (2007). Does dopamine dysfunction drive depression? *Acta Psychiatrica Scandinavica Supplement*, 433, 116-124.

[55] Ishrat, T., Parveen, K., Khan, M. M., Khuwaja, G., Khan, M. B., Yousuf, S., et al. (2009). Selenium prevents cognitive decline and oxidative damage in rat model of streptozotocin-induced experimental dementia of Alzheimer's type. *Brain Research*, 1281, 117-127.

[56] Cardoso, B. R., Ong, T. P., Jacob-Filho, W., Jaluul, O., Freitas, M. I., & Cozzolino, S. M. (2010). Nutritional status of selenium in Alzheimer's disease patients. *British Journal of Nutrition*, 103, 803-806.

[57] Finley, J. W., & Penland, J. G. (1998). Adequacy or deprivation of dietary selenium in healthy men: clinical and psychological findings. *Journal of Trace Elements in Experimental Medicine*, 11, 1-27.

[58] Mokhber, N., Namjoo, M., Tara, F., Boskabadi, H., Rayman, M. P., Ghayour-Mobar-han, M., et al. (2011). Effect of supplementation with selenium on postpartum de-pression: a randomized doubleblind placebo-controlled trial. *Journal of Maternal-Fetal and Neonatal Medicine*, 24, 104-108.

[59] ISAAC Steering Committee. (1998). Worldwide variation in prevalence of symptoms of asthma, allergic rhinoconjunctivitis, and atopic eczema: ISAAC. *Lancet*, 351, 1225-1232.

[60] Allan, K., & Devereux, G. (2011). Diet and asthma: nutrition implication from pre-vention to treatment. *Journal of the American Dietetic Association*, 111, 258-268.

[61] Comhair, S., Bhathena, P., Farver, C., Thunnissen, F., & Srzurum, S. (2001). Extracel-lular glutathione peroxidase induction in asthmatic lungs: evidence for redox regula-tion of expression in human airway epithelial cells. *FASEB Journal*, 15, 70-78.

[62] Horváthovà, M., Jahnová, E., & Gazdik, F. (1999). Effect of selenium supplementation in asthmatic subjects on the expression of endothelial cell adhesion molecules in cul-ture. *Biological Trace Element Research*, 69, 15-26.

[63] Hasselmark, L., Malmgren, R., Zetterstrom, O., & Unge, G. (1993). Selenium supple-mentation in intrinsic asthma. *Allergy*, 48, 30-36.

[64] Picado, C., Deulofeu, R., Lleonart, R., et al. (2001). Dietary micronutrients/antioxi-dants and their relationship with bronchial asthma severity. *Allergy*, 56, 43-49.

[65] Shaheen, S., Sterne, J., Thompson, R., Songhurst, C., Margetts, B., & Burney, P. (2001). Dietary antioxidants and asthma in adults. Population-based case-control study. *American Journal of Respiratory and Critical Care Medicine*, 164, 1823-1828.

[66] Burney, P., Potts, J., Makowska, J., et al. (2008). A case-control study of the relation between plasma selenium and asthma in European populations: a GAL2EN project. [see comment][erratum appears in Allergy. 2008; 63 1647.], *Allergy*, 63, 865-871.

[67] Thomson, C. D., Wickens, K., Miller, J., Ingham, T., Lampshire, P., Epton, Town. G. I., Pattemore, P., & Crane, J. (2012). Selenium status and allergic disease in a cohort of New Zealand children. *Clinical & Experimental Allergy*, 42, 560-567.

[68] Nogueira, C. W., Zeni, G., & Rocha, J. B. (2004). Organoselenium and organotelluri-um compounds: toxicology and pharmacology. *Chemical Reviews*, 104, 6255-6285.

[69] Silva Avila, D., Benedetto, A., Au, C., Manarin, F., Erikson, K., Antunes Soares, F., Teixeira Rocha, J. B., & Aschner, M. (2012). Organotellurium and organoselenium compounds attenuate Mn-induced toxicity in Caenorhabditis elegans by preventing oxidative stress. *Free Radical Biology & Medicine*, 52, 1903-1910.

[70] Fetoui, H., Mahjoubi-Samet, A., Jamoussi, K., Ellouze, F., Guermazi, F., & Zeghal, N. (2006). Energy restriction in pregnant and lactating rats lowers bone mass of their progeny. *Nutrition Research*, 26, 421-426.

[71] Moreno-Reyes, R., Egrise, D., Boelaert, M., Goldman, S., & Meuris, S. (2006). Iodine deficiency mitigates growth retardation and osteopenia in selenium-deficient rats. *Journal of Nutrition*, 136, 595-600.

[72] Ren, F. L., Guo, X., Zhang, R. L., Wang, Sh. J., Zuo, H., Zhang, Z. T., et al. (2007). Effects of selenium and iodine deficiency on bone, cartilage growth plate and chondrocyte differentiation in two generations of rats. *Osteoarthritis and Cartilage*, 15, 1171-1177.

[73] Pahuja, D. N., & De Luca, H. F. (1982). Thyroid hormone and vitamin D metabolism in the rat. *Archives of Biochemistry and Biophysics*, 213, 293-298.

[74] Amaraa, I. B., Troudia, A., Soudania, N., Guermazib, F., & Zeghala, N. (2012). Toxicity of methimazole on femoral bone in suckling rats: Alleviation by selenium. *Experimental and Toxicologic Pathology*, 64, 187-195.

[75] UNAIDS. (2008). Report on the global AIDS epidemic. *Geneva: UNAIDS*.

[76] Fawzi, W. W., Msamanga, G. I., Spiegelman, D., et al. (2004). A randomized trial of multivitamin supplements and HIV disease progression and mortality. *The New England Journal of Medicine*, 351, 23-32.

[77] Diamond, A. M., Hu, J. Y., & Mansur, D. B. (2001). Glutathione peroxidase and viral replication: Implications for viral evolution and chemoprevention. *Biofactors*, 14, 205-210.

[78] Pace, G. W., & Leaf, C. D. (1995). The role of oxidative stress in HIV disease. *Free Radical Biology & Medicine*, 19, 523-528.

[79] Stone, C. A., Kawai, K., Kupka, R., & Fawzi, W. W. (2010). Role of selenium in HIV infection. . Nutrition Reviews , 68, 671-681.

[80] St Germain, D. L., Galton, V. A., & Hernandez, A. (2009). Minireview: Defining the roles of the iodothyronine deiodinases: current concepts and challenges. *Endocrinology*, 150, 1097-1107.

[81] Zhang, S., Rocourt, C., & Cheng, W-H. (2010). Selenoproteins and the aging brain. *Mechanisms of Ageing and Development*, 13, 253-260.

[82] Schweizer, U., Brauer, A. U., Kohrle, J., Nitsch, R., & Savaskan, N. E. (2004). Selenium and brain function: a poorly recognized liaison. *Brain Research Reviews*, 45, 164-178.

[83] Hock, A., Demmel, U., Schicha, H., Kasperek, K., & Feinendegen, L. E. (1975). Trace element concentration in human brain. Activation analysis of cobalt, iron, rubidium, selenium, zinc, chromium, silver, cesium, antimony and scandium. *Brain*, 98, 49-64.

[84] Prohaska, J. R., & Ganther, H. E. (1976). Selenium and glutathione peroxidase in developing rat brain. *Journal of Neurochemistry*, 27, 1379-1387.

[85] Trapp, G. A., & Millam, J. (1975). The distribution of ^{75}Se in brains of selenium-deficient rats. *Journal of Neurochemistry*, 24, 593-595.

[86] Nakayama, A., Hill, K. E., Austin, L. M., Motley, A. K., & Burk, R. F. (2007). All regions of mouse brain are dependent on selenoprotein P for maintenance of selenium. *Journal of Nutrition*, 137, 690-693.

[87] Kyriakopoulos, A., Rothlein, D., Pfeifer, H., Bertelsmann, H., Kappler, S., & Behne, D. (2000). Detection of small selenium-containing proteins in tissues of the rat. *Journal of Trace Elements in Medicine and Biology*, 14, 179-183.

[88] Hawkes, W. C., & Hornbostel, L. (1996). Effects of dietary selenium on mood in healthy men living in a metabolic research unit. *Biological Psychiatry*, 39, 121-128.

[89] Ashrafi, M. R., Shabanian, R., Abbaskhanian, A., Nasirian, A., Ghofrani, M., Mohammadi, M., Zamani, G. R., Kayhanidoost, Z., Ebrahimi, S., & Pourpak, Z. (2007). Selenium and Intractable Epilepsy: Is There Any Correlation? *Pediatric Neurology*, 36, 25-29.

[90] Rayman, M. P. (2000). The importance of selenium to human health. *Lancet*, 356, 233-241.

[91] Chen, J., & Berry, M. J. (2003). Selenium and selenoproteins in the brain and brain diseases. *Journal of Neurochemistry*, 86, 1-12.

[92] Berry, M. J., Banu, L., & Larsen, P. R. (1991). Type I iodothyronine deiodinase is a selenocysteine-containing enzyme. *Nature*, 349, 438-440.

[93] Hill, K. E., McCollum, G. W., Boeglin, M. E., & Burk, R. F. (1997). Thioredoxin reductase activity is decreased in selenium deficiency. *Biochemical and Biophysical Research Communications*, 234, 293-295.

[94] Duntas, L. H. (2010). Selenium and the Thyroid: A Close-Knit Connection. *The Journal of Clinical Endocrinology & Metabolism*, 95, 5180-5188.

[95] Hatfield, D. L., Yoo, M.-H., Carlson, B. A., & Gladyshev, V. N. (2009). Selenoproteins that function in cancer prevention and promotion. *Biochimica et Biophysica Acta*, 1790, 1541-1545.

[96] Steinbrenner, H., & Sies, H. (2009). Protection against reactive oxygen species by selenoproteins. *Biochimica et Biophysica Acta*, 1790, 1478-1485.

[97] Ip, C. (1998). Lessons from basic research in selenium and cancer prevention. *Journal of Nutrition*, 128, 1845-1849.

[98] Shamberger, R. J. (1970). Relationship of selenium to cancer: inhibitory effect of selenium on carcinogenesis. *Journal of the National Cancer Institute*, 44, 931-936.

[99] Clark, L. C., Combs, G. F., & Turnbull, B. W. (1996). Effect of selenium supplementation for cancer prevention in patients with carcinoma of the skin: a randomized clinical trial. *Journal of the American Medical Association*, 279, 1975-1982.

[100] Bjelakovic, G., Nikolova, D., Simonetti, R. G., & Gluud, C. (2008). Systematic review: primary and secondary prevention of gastrointestinal cancers with antioxidant supplements. *Alimentary Pharmacology & Therapeutics*, 28, 689-703.

[101] Brinkman, M., Buntinx, F., Muls, E., & Zeegers, M. P. (2006). Use of selenium in chemo- prevention of bladder cancer. *The Lancet Oncology*, 7, 766-774.

[102] Lu, J., & Jiang, C. (2005). Selenium and cancer chemoprevention: hypotheses integrating the actions of selenoproteins and selenium metabolites in epithelial and non-epithelial target cells. *Antioxidants & Redox Signaling*, 7, 1715-1727.

[103] Waters, D. J., Shen, S., Glickman, L. T., Cooley, D. M., Bostwick, D. G., Qian, J., et al. (2005). Prostate cancer risk and DNA damage: translational significance of selenium supple- mentation in a canine model. *Carcinogenesis*, 26, 1256-1562.

[104] Battin, E. E., Perron, N. R., & Brumaghim, J. L. (2006). The central role of metal coordination in selenium antioxidant acitivity. *Inorganic Chemistry*, 45, 499-501.

[105] Rayman, M. P. (2009). Selenoproteins and human health: Insights from epidemiological data. *Biochimica et Biophysica Acta*, 1790, 1533-1540.

[106] Bleys, J., Navas-Acien, A., & Guallar, E. (2007). Serum selenium and diabetes in US adults. *Diabetes Care*, 30, 829-834.

[107] Akbaraly, T. N., Arnaud, J., Rayman, M. P., Hininger-Favier, I., Roussel, A. M., Berr, C., & Fontbonne, A. (2010). Plasma selenium and risk of dysglycemia in an elderly French population: results from the prospective Epidemiology of Vascular Ageing Study. *Nutrition & Metabolism*, 7, 21-27.

[108] Bleys, J., Navas-Acien, A., & Guallar, E. (2007). Selenium and diabetes: more bad news for supplements. *Annals of Internal Medicine*, 147, 271-272.

[109] Faure, P., Ramon, O., Favier, A., & Halimi, S. (2004). Selenium supplementation decreases nuclear factor-kappa B activity in peripheral blood mononuclear cells from type 2 diabetic patients. *European Journal of Clinical Investigation*, 34, 475-481.

[110] Stranges, S., Marshall, T. R., & Natarajan, R. (2007). Effects of long-term selenium supplementation on the incidence of type 2 diabetes: a randomized trial. *Annals of Internal Medicine*, 147, 217-223.

[111] Agarwal, A., & Prabhakaran, S. A. (2005). Mechanism, measurement and prevention of oxidative stress in male reproductive physiology. *Indian Journal of Experimental Biology*, 43, 963-974.

[112] Hansen, J. C., & Degachi, Y. (1996). Selenium and fertility in animals and men: a review, *Acta Veterinaria Scandinavica*, 37, 19-25.

[113] Wantanobe, T., & Endo, A. (1991). Effects of selenium deficiency on spermmorphology and spermatocyte chromosomes in mice. *Mutation Research*, 262, 93-96.

[114] Olson, G. E., Winfrey, V. P., Hill, K. E., & Burk, R. F. (2004). Sequential development of flagellar defects in spermatids and epididymal spermatozoa of selenium deficient rats. *Reproduction*, 127, 335-341.

[115] Lenzi, A., Gandini, L., Lombardo, F., Picardo, M., Maresca, V., Panfili, E., et al. (2002). Polyunsaturated fatty acids of germ cell membranes, glutathione and blutathione dependent enzyme-PHGPx: from basic to clinic. *Contraception*, 65, 301-305.

[116] Hawkes, W. C., Alkan, Z., & Wong, K. (2009). Selenium supplementation does not affect testicular selenium status and semen quality in North American men. *Journal of Andrology*, 30, 525-533.

[117] Seaton, A., Godden, D. J., & Brown, K. (1994). Increase in asthma: a more toxic environment or a more susceptible population. *Thorax*, 49, 171-174.

[118] University of Maryland. Asthma [Online]. (2011). cited on August 10, 2011]. Available from URL: http://www.umm.edu/altmed/articles/asthma-000015.htm Citation Date= July 4, 2012].

[119] Kloner, A. R., Przyklenk, K., & Whittaker, P. (1989). Deleterious effects of oxygen radicals in ischaemia-reperfusion: resolved and unresolved issue. *Circulation*, 80, 1115-1127.

[120] Prithviraj, T., & Misra, K. P. (2000). Reversal of atherosclerosis-fact or fiction? *Cardiology Today*, 4, 97-100.

[121] Sattler, W., Maiorino, M., & Stocker, R. (1994). Reduction of HDL and LDL associated cholestrylester and phospholipids hydroperoxides by phospholipids hydroperoxide-glutathione peroxidase and ebselen (Pz 51). *Archives of Biochemistry and Biophysics*, 309, 224.

[122] Ricetti, M. M., Guidi, G. C., Tecchio, C., et al. (1999). Effects of sodium selenite on in vitro interactions between platelets and endotelial cells. *International Journal of Clinical and Laboratory Research*, 29, 80-82.

[123] Kharb, S. (2003). Low blood glutathione levels in acute myocardial infarction. *Indian Journal of Medical Sciences*, 57, 335-337.

[124] Riaz, M., & Mehmood, K. T. (2012). Selenium in human health and disease: a review. *Journal of Postgraduate Medical Institute*, 26, 120-133.

[125] Suleyman, Ö., Mustafa, N., Mesut, Ç., Vedat, B., & Flores-Arce, M. F. (2011). Effects of Different Medical Treatments on Serum Copper, Selenium and Zinc Levels in Patients with Rheumatoid Arthritis. *Biological Trace Element Research*, 142, 447-455.

[126] Peretz, A., Siderova, V., & Neve, J. (2001). Selenium supplementation in rheumatoid arthritis investigated in a double blind, placebo- controlled trial. *Scandinavian Journal of Rheumatology*, 30, 208-212.

[127] Cunningham, F. G., & Lindheimer, M. D. (1992). Hypertension in Pregnancy: Current concepts. *The New England Journal of Medicine*, 326, 927-932.

[128] Sharma, J. B. (2001). Benefits of Selenium during pregnancy. *Obstetrics and Gynaecology*, 6, 459-462.

[129] Barrington, J. W., Lindsay, P., Names, D., Smith, S., & Robert, A. (1996). Selenium deficiency and miscarriage: a possible link? *British Journal of Obstetrics and Gynecology*, 2, 130-132.

[130] Tinggi, U. (2008). Selenium: its role as antioxidant in human health. *Environmental Health and Preventive Medicine*, 13, 102-108.

[131] Beck, M. A. (2001). Antioxidants and viral infections: host immune response and viral pathogenicity. *Journal of the American College of Nutrition*, 20, 384S-388S.

[132] Beck, M. (2006). Selenium, viral infections. Hatfield DL, Berry MJ, Gladyshev VN, editors, *Selenium: its molecular biology and role in human health*, New York: Springer, 287-298.

[133] Broome, C. S., McArdle, F., Kyle, J. A., Andrews, F., Lowe, N. M., Hart, C. A., et al. (2004). An increase in selenium intake improves immune function and poliovirus handling in adults with marginal selenium status. *The American Journal of Clinical Nutrition*, 80, 154-162.

[134] Baum, M. K., & Campa, A. (2006). Role of selenium in HIV/AIDS. In: Hatfield DL, Berry MJ, Gladyshev VN, editors. *Selenium- its molecular biology and role in human health*, New York: Springer, 299-310.

[135] McKenzie, R. C., Beckett, G. J., & Arthur, J. R. (2006). Effects of selenium on immunity and aging. In: Hatfield DL, Berry MJ, Gladyshev VN editors. *Selenium-its molecular biology and role in human health*, New York: Springer, 287-298.

[136] Sheweita, S. A., & Koshhal, K. (2007). Calcium metabolism and oxidative stress in bone fractures: role of antioxidants. *Current Drug Metabolism*, 8, 519-525.

[137] Finkel, T., & Holbrook, N. J. (2000). Oxidants, oxidative stress and biology of ageing. *Nature*, 408, 147-239.

[138] Cetinus, E., Kilinc, M., Uzel, M., Inanc, F., Kurutas, E. B., Bilgic, E., et al. (2005). Does long-term ischemia affect the oxidant status during fracture healing? *Archives of Orthopedic and Trauma Surgery*, 125, 376-380.

[139] Sandukji, A., Al-Sawaf, H., Mohamadin, A., Alrashidi, Y., & Sheweita, S. A. (2010). Oxidative stress and bone markers in plasma of patients with long-bone fixative surgery: Role of antioxidants. *Human and Experimental Toxicology*, 30, 435-442.

[140] Azzi, A., Aratri, E., Boscoboinik, D., Clement, S., Ozer, N. K., Ricciarelli, R., & Spycher, S. (1998). Molecular basis of α-tocopherol control of smooth muscle cell proliferation. *Biofactors*, 7, 3-14.

[141] Brigelius-Flohe', R., & Galli, F. (2010). Vitamin E: A vitamin still awaiting the detection of its biological function. *Mol. Nutr. Food Res.*, 54, 583-587.

[142] Cicek, N., Eryilmaz, O. G., Sarikaya, E., Gulerman, C., & Genc, Y. (2012). Vitamin E effect on controlled ovarian stimulation of unexplained infertile women. *Journal of Assisted Reproduction and Genetics*, 29, 325-328.

[143] Khanna, S., Parinandi, N. L., Kotha, S. R., Roy, S., Rink, C., Bibus, D., & Sen, C. K. (2010). Nanomolar vitamin E alpha-tocotrienol inhibits glutamate-induced activation of phospholipase A2 and causes neuroprotection. *Journal of Neurochemistry*, 112, 1249-1260.

[144] Reiter, E., Jiang, Q., & Christen, S. (2007). Anti-inflammatory properties of alpha- and gamma-tocopherol. *Molecular Aspects of Medicine*, 28, 668-691.

[145] Naito, Y., Shimozawa, M., Kuroda, M., Nakabe, N., Manabe, H., Katada, K., Kokura, S., Ichikawa, H., Yoshida, N., Noguchi, N., & Yoshikawa, T. (2005). Tocotrienols reduce 25-hydroxycholesterol-induced monocyteendothelial cell interaction by inhibiting the surface expression of adhesion molecules. *Atherosclerosis*, 180, 19-25.

[146] Burlakova, E. B., Krashakov, S. A., & Khrapova, N. G. (1998). The role of tocopherols in biomembrane lipid peroxidation. *Membrane and Cell Biology*, 12, 173-211.

[147] Coquette, A., Vray, B., & Vanderpas, J. (1986). Role of vitamin E in the protection of the resident macrophage membrane against oxidative damage. Archives Internationales de Physiologie. de Biochimie et de Biophysique, 94 ., 29S-34S.

[148] Pratico, D., Tangirala, R. K., Rader, D. J., Rokach, J., & FitzGerald, G. A. (1998). Vitamin E suppresses isoprostane generation in vivo and reduces atherosclerosis in ApoE-deficient mice. *Nature Medicine*, 4, 1189-1192.

[149] Moriguchi, S., & Muraga, M. (2000). Vitamin E and immunity. *Vitamins & Hormones*, 59, 305-336.

[150] Wu, D., et al. (2000). In vitro supplementation with different tocopherol homologues can affect the function of immune cells in old mice. *Free Radical Biology & Medicine*, 28, 643-651.

[151] Palamanda, J. R., & Kehrer, J. P. (1993). Involvement of vitamin E and protein thiols in the inhibition of microsomal lipid peroxidation by glutathione. *Lipids*, 28, 427-431.

[152] Agarwal, A., & Sekhon, L. H. (2011). Oxidative stress and antioxidants for idiopathic oligoasthenoteratospermia: is it justified? *Indian Journal of Urology*, 27(1), 74-85.

[153] Bouayed, J., & Bohn, T. (2010). Exogenous antioxidants-double-edged swords in cellular redox state. Health beneficial effects at physiologic doses versus deleterious effects at high doses. *Oxidative Medicine and Cellular Longevity*, 3, 228-237.

[154] Chappell, L. C., Seed, P. T., Briley, A. L., et al. (1999). Effect of antioxidants on the occurrence of pre-eclampsia in women at increased risk: a randomized trial. *Lancet*, 354, 810-816.

[155] Fairfield, K. M., & Fletcher, R. H. (2002). Vitamins for chronic disease prevention in adults: scientific review. *Journal of the American Medical Association*, 287, 3116-3126.

[156] Forstrom, J. W., Zakowski, J. J., & Tappel, A. L. (1978). Identification of the catalytic site of rat liver glutathione peroxidase as selenocysteine. *Biochemistry*, 17, 2639-2644.

[157] Tamura, T., & Stadtman, T. C. (1996). A new selenoprotein from human lung adeno-
       carcinoma cells: purification, properties, and thioredoxin reductase activity. *Proceed-
       ings of the National Academy of Sciences*, 93, 1006-1011.

[158] Kiremidjian-Schumacher, L., Roy, M., Wishe, H. I., Cohen, M. W., & Stotzky, G.
       (1994). Supple- mentation with selenium and human immune cell functions. II. Effect
       on cytotoxic lymphocytes and natural killer cells. *Biological Trace Element Research*, 41,
       115-127.

[159] Baker, S. S., & Cohen, H. J. (1983). Altered oxidative metabolism in selenium-defi-
       cient rat granulocytes. *The Journal of Immunology*, 130, 2856-2860.

[160] Sakaguchi, S., et al. (2000). Roles of selenium in endotoxin-induced lipid peroxidation
       in the rats liver and in nitric oxide production in J774A. 1 cells. Toxicology Letters,
       118, 69-77.

[161] Kim, S. H., Johnson, V. J., Shin, T. Y., & Sharma, R. P. (2004). Selenium attenuates lip-
       opoly- saccharide-induced oxidative stress responses through modulation of p38
       MAPK and NF-kappaB signaling pathways. *Experimental Biology and Medicine*, 229,
       203-213.

[162] Maehira, F., Miyagi, I., & Eguchi, Y. (2003). Selenium regulates transcription factor
       NF-kappaB activation during the acute phase reaction. *Clinica Chimica Acta*, 334,
       163-171.

[163] Beck, M. A., & Matthews, C. C. (2000). Micronutrients and host resistance to viral in-
       fection. *Proceedings of the Nutrition Society*, 59, 581-585.

[164] Ip, C., & Medina, D. (1987). Current concepts of selenium and mammary tumorigen-
       esis. Medina D, Kidwell W, Heppner GH, Anderson E, editors, *Cellular and molecular
       biology of mammary cancer*, New York: Plenum press, 479.

[165] Ip, C., & Daniel, F. B. (1985). Effects of selenium on 7, 12 di-methyl benzanthracene
       induced mammary carcinogenesis and DNA adduct formation. *Cancer Research*, 45,
       61-68.

[166] Lotan, Y., Goodman, P. J., Youssef, R. F., Svatek, R. S., Shariat, S. F., Tangen, C. M.,
       Thompson, I. M. Jr., & Klein, E. A. (2012). Evaluation of Vitamin E and Selenium
       Supplementation for the Prevention of Bladder Cancer in SWOG Coordinated SE-
       LECT. *The Journal of Urology*, 187, 2005-2010.

[167] Jemal, A., Siegel, R., Xu, J., et al. (2010). Cancer statistics. *CA A Cancer Journal of Clini-
       cians*, 60, 277-300.

[168] Samanic, C., Kogevinas, M., Dosemeci, M., Malats, N., Real, F. X., Garcia-Closas, M.,
       Serra, C., Carrato, A., et al. (2006). Smoking and bladder cancer in Spain: effects of
       tobacco type, timing, environmental tobacco smoke, and gender. *Cancer Epidemiology,
       Biomarkers and Prevention*, 15, 1348-1354.

[169] Botteman, M. F., Pashos, C. L., Redaelli, A., et al. (2003). The health economics of bladder cancer: a Comprehensive review of the published literature. *Pharmacoeconomics*, 21, 1315-1330.

[170] Amaral, A. F. S., Cantor, K. P., Silverman, D. T., & Malats, N. (2010). Selenium and bladder cancer risk: a meta-analysis. *Cancer Epidemiol Biomarkers Prev*, 19, 2407-2415.

[171] Bardia, A., Tleyjeh, I. M., Cerhan, J. R., et al. (2008). Efficacy of antioxidant supplementation in reducing primary cancer incidence and mortality: systematic review and meta-analysis. *Mayo Clinic Proceedings*, 83, 23-34.

[172] Lippman, S. M., et al. (2009). Effect of selenium and vitamin E on risk prostate cancer and other cancers: the Selenium and Vitamin E Cancer Prevention Trial (SELECT). *Journal of American Medical Association*, 301, 39-51.

[173] Ledesma, M. C., Jung-Hynes, B., Schmit, T. L., Kumar, R., Mukhtar, H., & Ahmad, N. (2011). Selenium and vitamin E for prostate cancer: Post-SELECT (Selenium and Vitamin E Cancer Prevention Trial) status. *Molecular Medicine*, 17, 134-143.

[174] Beck, M. A. (2007). Selenium and vitamin E Status: Impact on viral pathogenicity. *The Journal of Nutrition*, 137, 1338-1340.

[175] Jiang, L., Yang, K-h., Tian, J-h., et al. (2010). Efficacy of Antioxidant Vitamins and Selenium Supplement in Prostate Cancer Prevention: A Meta-Analysis of Randomized Controlled Trials. *Nutrition and Cancer*, 62, 719-727.

[176] Jin, X., Hidiroglou, N., Lok, E., Taylor, M., Kapal, K., et al. (2012). Dietary Selenium (Se) and Vitamin E (VE) Supplementation Modulated Methylmercury-Mediated Changes in Markers of Cardiovascular Diseases. *Rats Cardiovascular Toxicology*, 12, 10-24.

[177] El-Desoky, G., Abdelreheem, M., AL-Othman, A., Alothman, Z., Mahmoud, M., & Yusuf, K. (2012). Potential hepatoprotective effects of vitamin E and selenium on hepatotoxicity induced by malathion in rats. *African Journal of Pharmacy and Pharmacology*, 6, 806-813.

Chapter 3

The Role of Natural Antioxidants in Cancer Disease

Carmen Valadez-Vega, Luis Delgado-Olivares,
José A. Morales González, Ernesto Alanís García,
José Roberto Villagomez Ibarra,
Esther Ramírez Moreno, Manuel Sánchez Gutiérrez,
María Teresa Sumaya Martínez,
Zuñiga Pérez Clara and Zuli Calderón Ramos

Additional information is available at the end of the chapter

1. Introduction

Cell oxidation can lead to the onset and development of a wide range of diseases including Alzheimer and Parkinson, the pathologies caused by diabetes, rheumatoid arthritis, neuro-degeneration in motor neuron diseases, and cancer. Reactive species (RS) of various types are powerful oxidizing agents, capable of damaging DNA and other biomolecules. Increased formation of RS can promote the development of malignancy, 'normal' rates of RS generation may account for the increased risk of cancer development.

Oxidants and free radicals are inevitably produced during the majority of physiological and metabolic processes and the human body has defensive antioxidant mechanisms; these mechanisms vary according to cell and tissue type and may act antagonistically or synergistically. They include natural enzymes like Superoxide dismutase (SOD), Catalase (CAT), and Gluta-thione peroxidase (GPx), as well as antioxidants such as vitamins, carotenoids, polyphenols, and other natural antioxidants, which have attracted great interest in recent years.

There has been a great deal of interest of late in the role of complementary and alternative drugs for the treatment of various acute and chronic diseases. Among the several classes of phytochemicals, interest has focused on the anti-inflammatory and antioxidant properties of the polyphenols that are found in various botanical agents. Plant vegetables and spices used in folk and traditional medicine have gained wide acceptance as one of the main sources of prophylactic and chemopreventive drug discoveries and development.

Recently, researches on medicinal plants has drawn global attention; large bodies of evidence have accumulated to demonstrate the promising potential of medicinal plants used in various traditional, complementary, and alternate treatment systems of human diseases. The plants are rich in a wide variety of secondary metabolites, such as tannins, terpenoids, alkaloids, flavonoids, etc., which have been screened *in vivo* and *in vitro* and have indicated antioxidant and anticarcinogenic properties and which are used to developed drugs or dietary supplements.

Evidence suggests that the plant kingdom is considered a good candidate for chemoprevention and cancer therapy due to the high concentration and wide variety of antioxidants such as resveratrol, genestein, beicalein, vitamin A, vitamin C, polyphenols, (–)–Epigallocatechin 3-gallate, flavonoids, polyphenols, gallic acid, glycosides, verbascoside, calceorioside, epicatechin, quercetin, curcumin, lovastatin, and many other types of compounds with the capability to inhibit the cell proliferation of different cancer cells *in vitro* and *in vitro*, such as colon cancer (HT-29, SW48, HCT116), breast (MCF7, MDA), cervix (HeLa, SiHa, Ca-Ski, C33-A), liver (Hep G2), skin (A 431), fibroblasts (3T3 SV40), and many other malignant cells; studies have indicated that antioxidants can be employed efficiently as chemopreventives and as effective inhibitors of cell proliferation, promoting cell apoptosis, and increasing detoxification enzymes, and inhibiting gene expression and scavenger Reactive oxygen species (ROS). Thus, many researchers are working with different types of natural antioxidants with the aim of finding those with the greatest capacity to inhibit the development of cancer both *in vitro* as well as *in vivo*, because these compounds have exhibited high potential for use not only in the treatment of this disease, but they also act as good chemoprotective agents.

2. Antioxidants

The production of ROS during metabolism is an inevitable phenomenon associated with the process of aerobic metabolism; on the other hand, we are exposed at all times to several exogenous sources of oxidant molecules, for example, environmental and pollutant factors and many dietary compounds, which increase their levels. ROS participate in different cellular processes; their intracellular levels are relatively low. However, because ROS are highly toxic when their concentration increases, the phenomenon denominated Oxidative stress (OS) is produced [123], which can injure various cellular biomolecules, causing serious damage to tissues and organs and resulting in chronic diseases [24]. Oxidative damage can be prevented by antioxidants, which are present within the cell at low concentrations compared with oxidant molecules [141, 50].

Antioxidants are capable of donating electrons to stabilize ROS and to inhibit their detrimental effects, including both endogenous (synthesized by the body itself) and exogenous molecules (those from external sources to the body) [141]. Endogenous antioxidants include Superoxide dismutase (SOD), which catalyzes the dismutation reaction of superoxide ($O2\bullet^-$) into hydrogen peroxide (H_2O_2), which is in turn transformed into oxygen and water for the Catalase (CT), and in addition Glutathione peroxidase (GPx) can catalyze its reduction; however, if in the presence of transition metals such as iron, H_2O_2, by means of the Fenton reaction, can produce the hydroxyl radical (OH•−); wich is of more reactive the ROS, capable to produce the majority of oxidative damage [24]. On the other hand, exogenous antioxidants can be from animal and plant sources; however, those of plant origin are of great interest because they can contain major antioxidant activity [19]. Different reports show that persons with a high intake of a diet rich in fruit and vegetables have an important risk reduction of developing cancer, mainly due to their antioxidant content [70]. Among the vegetable antioxidants are vitamins E and C, and ß-carotene, which are associated with diminished cardiovascular disease and a decreased risk of any cancer [48]. In particular, ß-carotene and vitamin E can reduce the risk of breast cancer, vitamin C, ß-carotene, and lutein/zeaxanthin possess a protector effect against ovarian cancer, and vitamin C, ß-carotene, and rivoflavin prevent colorectal cancer [70], while flavonoids such as plant phenolics and wine phenolics can inhibit lipid peroxidation and lipoxygenase enzymes. In addition, any microelement, such as Se, Zn, Mn, and Cu, can exhibit antioxidant activity [48, 24].

In recent years, interest has grown in the use of natural antioxidants for the prevention or treatment of different diseases related with OS; however despite the widespread information of the beneficial effects of antioxidants in the prevention of cancer, their use remains questionable, because different reports have shown that reducing the levels of ROS may have counterproductive effects because due to raising the risk of cancer; the latter may be due to that ROS can produce apoptosis in malignant cells [38, 101].

3. Molecular Studies of Natural Antioxidants

Different types of natural antioxidants are present in fruit and vegetables; they have synergistic interactions that are important due to their activity and regenerative potential. For example, ascorbate can regenerate into α-tocopherol [53], and the ascorbate radical is regenerated into other antioxidants via the thiol redox cycle. Taken together, all of these interactions are known as the "antioxidant network".

Vitamin E is an antioxidant that penetrates rapidly through the skin and is incorporated into the cellular membranes, inhibiting lipid peroxidation; specifically, α-tocotrienol, the vitamin E isoform, demonstrates greatest protection. Additionally, vitamin E possesses antiproliferative properties that interfere in signal transduction and in inducing cell cycle arrest.

Tumor necrosis factor-alpha (TNF-α) is a cytokine that, under normal conditions, induces inflammation, tumor inhibition, and apoptotic cell death. However, when the former undergoes deregulation, it acts as a breast tumor promoter, enhancing the proliferation of chemically induced mammary tumors [113]. Phenolic antioxidants can block the increase of TNF-α at the transcriptional level in the nucleus, which suggests the molecular mechanism of phenolic antioxidants through control of cytokine induction [81].

4. Oxidative Stress and Diseases

The ROS, as the superoxide anion ($O_2\bullet^-$), hydrogen peroxide (H_2O_2), and the hydroxyl radical ($OH\bullet$), are produced during cell metabolism in the lysosomes, peroxisomes, endoplasmic reticulum in the process carried out to obtain energy such as Adenosine triphosphate (ATP) [108]. There are other sources of oxidant molecules, such as pollution, the environment, and certain foods. During recent years, it has been discovered that during aging, the mitochondria increase the levels of ROS production and antioxidant endogens are diminished [98, 13]. ROS play an important role in the physiological process; however, due to their toxicity, their levels must be controlled by the endogenous antioxidant system. But when ROS formation is increased, an imbalance is promoted between these and the antioxidant molecules; phenomenon known as Oxidative stress (OS) [123]. OS can cause oxidative damage of proteins, lipids, and nucleic acids, macromolecules involved in the cell function, membrane integrity, or in maintaining genetic information (nucleic acids) [44, 45, 65].

Proteins are responsible for different cell processes (enzymatic, hormonal, structural support). The oxidation of proteins produces disulfide crosslinks, nitration, or tyrosine residues, and carbonylation, resulting in the loss of the structure and function of proteins and fragmentation [11, 97]. But because the chaperones are susceptible to oxidative damage, allowing the accumulation of misfolding proteins and increasing their susceptibility to protease degradation [115], however, the proteasome also undergoes oxidation and its activity is diminished, which makes the aggregates accumulate in the cell wich have been associated with aging and various pathologies, such as cancer and neurodegenerative disorders, such as Parkinson, Huntington, and Alzheimer disease [98].

The brain is the organ with the highest oxygen consumption; it has high levels of fatty acids, iron, and low antioxidant defenses. This is an organ with major susceptibility to oxidative damage [141], producing neurodegeneration that results in different diseases such as Parkinson disease, Alzheimer disease, Down syndrome, autism, bipolar disorder, and epilepsy [23, 24], and the cognitive alteration known as Mild cognitive impairment (MCI), which is produced preferentially in regions of the brain involved in regulating cognition, contributing to the development of dementia [65]. Similar processes occur during aging, resulting in the genetic response of increasing levels of antioxidant enzymes and chaperone proteins [73]. Reduction of OS causes improvement of the long-term memory [102].

Polyunsaturated fatty acids (mainly compounds of the membranes) are susceptible to peroxidation, which affects the integrity of the membranes of organelles of the cell membrane and the respiratory chain, in turn affecting cell viability. Lipid peroxidation produces aldehydes such as 4-hydroxy-2 E-nonenal, which is toxic and is involved in alterations in Alzheimer disease and DNA damage, causing mutations associated with the development of cancer [38, 20].

Ribosomal RNA and transfer RNA constitute the majority of stable species of cellular RNA, which possess a greater oxidation rate than DNA. The major modification for oxidation into RNA comprises 8-hydroxyGuanine (8-oxoG), which under normal conditions is present three times more in non-ribosomal that in ribosomal RNA; however, when the cell is exposed to H_2O_2, the concentration of 8-oxoG in ribosomal RNA increases at the same levels in both RNA [97]. RNA oxidation can diminish the capacity of replacement oxidation of proteins [65, 44] and the inhibition of protein synthesis, cell cycle arrest, and cell death. Oxidation of RNA is involved in the development of cancer, viral infections, AIDS, hepatitis (VIH-1; HCV; 107, 148], and neurological diseases. It has been reported that each neurological disease, present a damage oxidative of RNA in a specific region on the brain, for example in Alzheimer disease, there are increased RNA oxidation in the hippocampus and cerebral neocortex, while in Parkinson disease, RNA oxidation is localized in the *sustancia nigra* [97].

On the other hand, high-fat diets induce obesity and insulin resistance, resulting in increased ROS production, which modifies sympathetic brain activity, which in turn contributes to the rise in blood pressure, increase in insulin resistance, and obesity [6]. Obesity is the principal factor in the development of the metabolic syndrome, due to that persons with obesity have deficient antioxidant defense and increased production of ROS [126, 30, 75], which leads to spoilage and subsequently cell death, resulting in tissue and organ damage, to tissues causing serious health problems such as insulin resistance [7], diabetes mellitus, and hypertension [82]. Moreover, in the metabolic syndrome, NAD(P)H oxidase, the major source of ROS in several tissues, is up-regulated, resulting in an increase of ROS production and the down-regulation of several antioxidant enzymes (SOD isoforms, GPx, and heme oxygenase) [114]. This enzyme, specifically in the type 4 isoform (NOX4), is implicated in the damage due to OS during cerebral ischemia [67].

The scientific literature has shown that oxidative stress is involved in the development of a wide range of disease, such as heart diseases, Hutchinson-Gilford syndrome or progeria, hypertensive brain injury, muscular dystrophy, multiple sclerosis, congenital cataract, retinal degeneration, retinopathy of the premature, autoimmune diseases, cardiovascular abnormalities, nephrological disorders, emphysema, stroke, rheumatoid arthritis, anemia, hepatitis, pancreatitis, aging, premature wrinkles and dry skin, endothelial dysfunction, and dermatitis, among others [83, 7, 137, 91, 23, 102].

However the most important damage caused by OS are the DNA modifications, which can result in permanent mutations, due to that oxidative damage also affects the proteins involved in repairing the harm or reducing the OS (the endogenous antioxidant); thus, oxidative damage to DNA can be the cause of the development of various diseases, such as cancer [13, 51].

5. Cancer

Cancer is unnatural cell growth, in which cells can lose their natural function and spread throughout the blood in the entire body. Breast cancer is the most commonly diagnosed cancer in industrialized countries and has the highest death toll [88]. OS is involved in the process of the development of cancer and tumors, due to that ROS can damage the macromolecules as lipids, which react with metals (such as free iron and copper) and produce aldehydes and synthesize malondialdehyde-inducing mutations [96] or cause breaks in the double chain, produce modifications in guanine and thymine bases, and sister chromatid exchanges [16], which can affect the activities of signal transduction, transcription factors, and gene tumor suppressors such as *p53*, which is a gene important in apoptosis and in cell cycle control. This inactivation can increase the expression of proto-oncogenes [96] which can produce major damage. Oxidative damage or genetic defects that result in some defective enzymes are incapable of repairing the mutations increase the incidence of age-dependent cancer [51].

On the other hand, treatments with anticancer drugs and radiation increase ROS and decrease antioxidants content, producing a state of severe oxidative stress and causing apoptosis, resulting in side effects [96], while persistent oxidative stress at sublethal levels can result in resistance to apoptosis [16].

Some microorganisms, as bacteria and viruses, are involved, via OS, in the process of the production of certain cancers such as, for example *Helicobacter pylori*, inducing gastric cancer and colon cancer through the production of SO•⁻ [96]. It has been proposed that lower antioxidant activity increases the risk of developing cancer; thus, ingestion of antioxidants can prevent cancerogenesis. However is not clear the decrease of antioxidants levels is not clear, in as much as in freshly cancerous tissue, MnSOD levels are elevated; therefore, some investigators have proposed that this antioxidant enzyme is involved in tumor invasion; thus, it is possible that antioxidants have a role as pro-oxidants. Another point to consider is that when the 8-oxodG level in DNA increases, cancer rates do not increase [96, 51]. However, OS is a factor for cancer and other diseases, but not the sole factor for diseases, because others, such as genetic factors (genetic predisposition) are involved.

6. Antioxidants and Cancer

Humans are constantly bombarded by exogenous factors such as Ultraviolet (UV) rays, tobacco smoke, and many others agents that cause OS. Such stress can also arise from the drugs that are employed in medical practice. On the other hand, under physiological conditions, normal aerobic metabolism gives rise to active and potentially dangerous oxidants in cells and tissues; these endogenous sources of OS include those derived from the activities of mitochondria or microsomes and peroxisomes in the electron transfer system and from the activities of the NADPH enzyme present in macrophages and neutrophils as a mechanism of protection against infection. Various reducing substances in the human body control the status of oxidation-reduction (redox), and a continuing imbalance in favor of oxidation causes several problems when it exceeds the capacity of such a control [96].

Otto Warburg was the first scientist to implicate oxygen in cancer [147] as far back as the 1920s. However, the underlying mechanism by which oxygen might contribute to the carcinogenic process was undetermined for many years. The discovery of superoxide dismutase in 1968 by [90] led to an explosion of research on the role of reactive oxygen in the pathologies of biological organisms. Reactive oxygen has been specifically connected with not only cancer, but also many other human diseases [5, 57]. For many years, research on OS focused primarily on determining how ROS damage cells through indiscriminate reactions with the macromolecular machinery of a cell, particularly lipids, proteins, and DNA. It is well known and in great detail the manner in which ROS react with lipids, leading to the peroxidation of biological membranes and resulting in necrotic lesions [43] and the way ROS react with the nucleotides of DNA, leading to potential mutations [17, 43, 139].

When produced in excess, ROS (some of which are free radicals) can seriously alter the structure of biological substrates such as proteins, lipids, lipoproteins, and Deoxyribonucleic acid (DNA). They possess a huge range of potential actions on cells, and one could easily envisage them as anti-cancer (e.g., by promoting cell-cycle stasis, senescence, apoptosis, necrosis or other types of cell death, and inhibiting angiogenesis), or as pro-cancer (promoting proliferation, invasiveness, angiogenesis, metastasis, and suppressing apoptosis).

Active oxygen may be involved in carcinogenesis through two possible mechanisms: induction of gene mutations that result from cell injury [34], and the effects on signal transduction and transcription factors. Which mechanism it follows depends on factors such as the type of active oxygen species involved and the intensity of stress [86]. Cellular targets affected by oxidative stress include DNA, phospholipids, proteins, and carbohydrates on the cell membrane. Oxidized and injured DNA has the potential to induce genetic mutation. That some telomere genes are highly susceptible to mutation in the presence of free radicals is now apparent, and it is known that tumor suppressor genes such as *p53* and cell cycle-related genes may undergo DNA damage. In addition, oxidized lipids react with metals to produce active substances (e.g., epoxides and aldehydes) or synthesize malondialdehyde, which has the potential to induce mutation. Active oxygen species act directly or indirectly via DNA damage on gene expression (DNA binding of transcription factors) and signaling at the cellular level.

Markers for OS can be divided into three categories:

1.   formation of modified molecules by free radical reactions;

2.   consumption or induction of antioxidant molecules or enzymes, and

3.   activation or inhibition of transcription factors.

Targets of free radicals include all types of molecules in the body. Among these, lipids, nucleic acids, and proteins are the major targets. Because free radicals are usually generated near membranes (cytoplasmic membrane, mitochondria, or endoplasmic reticulum), lipid peroxidation is the first reaction to occur. Lipid peroxidation products can be detected as classical Thiobarbituric acid (TBA)-reactive substances. Recently, the detection of 4-Hydroxy-2-nonenal (HNE) or Malondialdehyde (MDA) is favored due to their high specificity [32], aldehydes are end-products of lipid peroxidation but continue to be reactive with cell proteins [136].

Exposure to free radicals from a variety of sources has led organisms to develop a series of defense mechanisms that involve the following:

1.   preventative mechanisms;

2.   repair mechanisms;

3.   physical defenses, and

4. antioxidant defenses.

Enzymatic antioxidant defenses include Superoxide dismutase (SOD), Glutathione peroxidase (GPx), and Catalase (CAT). Non-enzymatic antioxidants are represented by ascorbic acid (vitamin C), α-tocopherol (vitamin E), Glutathione (GSH), carotenoids, flavonoids, tannins, triterpepenoids, saponins, glycosides, steroids, and other antioxidants [46]. Under normal conditions, there is a balance between both the activities and the intracellular levels of these antioxidants: this equilibrium is essential for the survival of organisms and their health

7. Antioxidants in Cancer Assays

Humans have evolved with antioxidant systems for protection against free radicals and ROS. These systems include some antioxidants produced in the body (endogenous) and others obtained from the diet (exogenous) [21]. The former include

1. enzymatic defenses, such as Se-glutathione peroxidase, catalase, and superoxide dismutase, which metabolize superoxide, hydrogen peroxide, and lipid peroxides, thus preventing the majority of the formation of toxic HO•, and

2. non-enzymatic defenses, such as glutathione, histidine peptides, the iron-binding transfer proteins and ferritin, and dihydrolipoic acid, reduced Coenzyme Q10, melatonin, urate, and plasma protein thiols, with the latter two accounting for the major contribution to the radical-trapping capacity of plasma.

The various defenses are complementary to each other because they act against different species in different cellular compartments. However, despite these defense antioxidants (able either to suppress free radical formation and chain initiation or to scavenge free radicals and chain propagation), some ROS escape to cause damage. Thus, the body's antioxidant system is also provided with repair antioxidants (able to repair damage) and based on proteases, lipases, transferases, and DNA repair enzymes [145, 103].

Owing to the incomplete efficiency of our endogenous defense systems and the existence of some physiopathological situations (cigarette smoke, air pollutants, UV radiation, a high, polyunsaturated fatty acid diet, inflammation, ischemia/reperfusion, etc.) in which ROS are produced in excess and at the wrong time and place, dietary antioxidants are required to diminish the cumulative effects of oxidative damage throughout the human lifespan [149, 47). Well known natural antioxidants derived from the diet, such as vitamins C, E, and A and the carotenoids, have been studied intensively [124]. In addition to these, antioxidants in plants might account for at least part of the health benefits associated with vegetable and fruit consumption [103].

The plants, vegetables, and spices used in folk and traditional medicine have gained wide acceptance as one of the main sources of prophylactic and chemopreventive drug discovery and development [85, 29].

Some reports indicate that the prevalence of use of complementary and alternative medicine by patients with cancer has been estimated at a range of 7–64% [3, 4, 58]. At present, many patients with cancer combine some forms of complementary and alternative therapy with their conventional therapies [4, 58]. A recent survey of patients at a comprehensive cancer center placed the use of vitamin and minerals at 62.6%; of these patients, 76.6% combined the use of vitamins and minerals with conventional chemotherapy [58, 27].

These types of patients employ complementary and alternative therapies for a variety of reasons [31, 14]: to improve quality of life (77%); to improve immune function (71%); to prolong life (62%), or to relieve symptoms (44%) related with their disease [31]. Only 37.5% of the patients surveyed expected complementary and alternative therapies to cure their disease. Whatever the reasons, alternative therapy use is on the rise and this includes the use of megavitamins, minerals, and cocktails of natural substances during chemotherapy administration; these cocktails include antioxidants such as the commonly consumed antioxidants vitamin E (mixed tocopherols and tocotrienols), vitamin C, β-carotene (natural mixed carotenoids), polyphenols, tannins, terpenoids, alkaloids, flavonoids, vitamin A, and many others. Controversy exists concerning the use of antioxidants with chemotherapy, but increasing evidence suggests a benefit when antioxidants are added to chemotherapy [111, 112, 106, 151, 117, 105, 22, 27].

It is widely accepted that diets rich in fruits and plants are rich sources of different types of antioxidants; phenolic compounds are the most studied of these and have been recognized to possess a wide range of properties including antioxidant, antibacterial, anti-inflammatory, hepatoprotective, and anticarcinogenic actions [3, 4, 63]. Many of the biological functions of flavonoid, phenolic, catechin, curcumin, resveratrol, and genistein compounds have been attributed to their free-radical scavenging, metal-ion chelating, and antioxidant activities [118, 152]. Antioxidant phenolic agents have been implicated in the mechanisms of chemoprevention, which refers to the use of chemical substances of natural or of synthetic origin to reverse, retard, or delay the multistage carcinogenic process [29].

It has been shown that dietary phytochemicals can interfere with each stage of the development of carcinogenesis [130, 93]. As in the case of direct antioxidant effects, dietary polyphenols are most likely to exert their chemopreventive effects on the gastrointestinal tract, where they are present at highest concentrations [52, 49, 84, 75]. Indeed, studies have shown that various polyphenol-rich fruits and vegetables are particularly effective in protecting against several types of cancer development [84, 75, 59]. Dietary polyphenols may exert their anticancer effects through several possible mechanisms, such as removal of carcinogenic agents, modulation of cancer cell signaling and antioxidant enzymatic activities, and induction of apoptosis as well as of cell cycle arrest. Some of these effects may be related, at least partly, with their antioxidant activities [59]. They may exert protective effects against cancer development, particularly in the gastrointestinal tract, where they will be at their highest concentration. In fact, many studies have shown that various polyphenol-rich fruits and vegetables are particularly effective in protecting against colon cancer development [84, 75].

At the cellular level, there is good evidence that polyphenols present in tea, red wine, cocoa, fruit juices, and olive oil; at some level, they are able to stimulate carcinogenesis and tumor development [93]. For example, they may interact with reactive intermediates [28] and activated carcinogens and mutagens [18], they may modulate the activity of the key proteins involved in controlling cell cycle progression [104], and they may influence the expression of many cancer-associated genes [142]. Perhaps most notably, the anticancer properties of green tea flavanols have been reported in animal models and in human cell lines (Takada *et al.*, 2002], as well as in human intervention studies [60]. On the other hand, green tea consumption has been proposed as significantly reducing the risk of cancer of the biliary tract [133], bladder [110], breast [74], and colon [72]. Many of the anti-cancer properties associated with green tea are thought to be mediated by the flavanol Epigallocatechin gallate (EGCG), which has been shown to induce apoptosis and inhibit cancer cell growth by altering the expression of cell cycle regulatory proteins and the activity of signaling proteins involved in cell proliferation, transformation, and metastasis [66]. In addition to flavonoids, phenolic alcohols, lignans, and secoiridoids (all found at high concentrations in olive oil) are also thought to induce anti-carcinogenic effects [99] and have been reported in large intestinal cancer cell models [79], in animals [10, 128], and in humans [99]. These effects may be mediated by the ability of olive oil phenolics to inhibit initiation, promotion, and metastasis in human colon adenocarcinoma cells [42, 55] and to down-regulate the expression of COX-2 and Bcl-2 proteins, which play a crucial role in colorectal carcinogenesis [79, 146].

In vivo studies have demonstrated that many natural compounds found in plants and fruits have the capability to inhibit many types of human and animal cancer. Vitamins such as C, E, and A have shown that they can diminish cervical, bladder, prostate, intestinal, skin, and other gastrointestinal cancer types and that they have the capability to inhibit ROS production in patients [36, 37, 89, 134, 131, 62, 127]. In addition, it was demonstrated that these vitamins can inhibit progression and pathogenesis in colorectal cancer [12]. In animal models, vitamins showed promise for chemopreventive agents against several types of gastrointestinal cancer [62].

With the use of a combination of vitamins, selenium, β-carotene, essential fatty acids, and coenzyme Q10 in patients with breast cancer, it was observed that during the study no patient died, no patient showed signs of further distant metastasis, quality of life improved, and six patients showed apparent partial remission [80]. Human studies demonstrated that consumption of total antioxidants in the diet (fruits and vegetables) is inversely associated with the risk of distal gastric cancer [87]. Antioxidants, especially polyphenols, have been found to be promising agents against cervical cancer, including induction of apoptosis, growth arrest, inhibition of DNA synthesis, and modulation of signal transduction pathway; additionally, polyphenols can interfere with each stage of carcinogenesis initiation, promotion, and progression for the prevention of cancer development [26].

Camelia sinensis tea, which contains a great quantity of polyphenols (epichatechin, (–)–epigallocatechin-3-gallate) is the most widely consumed beverage worldwide, and it was demonstrated that consumption of this beverage has shown to afford protection against chemical carcinogen-induced stomach, lung, esophagus, duodenum, pancreas, liver, breast,

and colon carcinogenesis in specific bioassay models. The properties of the tea's polyphenols make them effective chemopreventive agents against the initiation, promotion, and progression stages of multistage carcinogenesis [64]. Rosmanic acid had demonstrated to
possess potent anticancer and apoptotic effect in mouse-induced skin cancer [121], curcumin, (–)–epigallocatechin-3-gallate, and lovastatin in combination were able to suppress
esophageal cancer in mouse [154], and melatonin demonstrated diminishing the development and mortality of mouse implanted with murine hepatoma cells MN22a [39]. It was
demonstrated that beta-ionone, a precursor of carotenoids, ameliorated lung carcinogenesis;
the latter is attributed to the antiproliferative and antioxidant potential of beta-ionone
through free radical scavenging properties [9]. A-tocopherol showed down-regulation of the
expression of the stress-activated genes *PKC-α, c-Myc*, and *Lactate dehydrogenase A (LDHA)* in
cancerous mice, decreasing cancer cell proliferation [120]. It has been suggested that rosmanic acid suppresses oral carcinogenesis by stimulating the activities of detoxification enzymes, improving the status of lipid peroxidation and antioxidants, and down-regulating
the expression of *p53* and *bcl-2* during 7,12 dimethylbenz(a)anthracene-induced oral carcinogenesis in hamster [8]. In the same manner, the methanolic extract of fennel seed exhibited
an antitumoral affect by modulating lipid peroxidation and augmenting the antioxidant defense system in Ehrlich ascites carcinoma- bearing mice with or without exposure to radiation [94]. Silymarin, a natural flavonoid from the milk thistle seed, displayed
chemopreventive action against 1,2-dimethylhydrazine plus dextran sodium sulfate-induced inflammation associated with colon carcinogenesis [135]. Quercetin, a flavonoid
found in many natural foods, demonstrated to exert a direct oro-apoptotic affect on tumor
cells and can indeed block the growth of several human cancer-cell lines in different cell-cycle phases, which have been demonstrated in several animal models [41]. The methanolic
extract of *Indigofera cassioides* was evaluated in terms of their antitumor activity on Ehrlich
ascites carcinoma- bearing mice; the extract showed a potent antitumoral effect against tumor cells due its preventing lipid peroxidation and promoting the enzymatic antioxidant
defense system in animals [69]. Brucine, a natural plant alkaloid, was reported to possess cytotoxic and antiproliferative activities and also had showed to be a potential anti-metastatic
and -angiogenic agent [2].

An *in vitro* assay demonstrated that the mechanism's antioxidant action, according to Halliwell [52], can include the following:

1.   suppressing ROS formation either by inhibiting the enzymes or chelating the trace ele-
     ments involved in free radical production;

2.   scavenging ROS, and

3.   up-regulating or protecting antioxidant defenses.

Flavonoids have been identified as fulfilling the majority of the criteria previously described. Thus, their effects are two-fold as follows:

1.   Flavonoids inhibit the enzymes responsible for superoxide anion production, such as
     xanthine oxidase [54] and Protein kinase C (PKC) [140], and

2. Flavonoids have also shown to inhibit cyclo-oxygenase, lipoxygenase, microsomal mono-oxygenase, glutathione S-transferase, mitochondrial succinoxidase, and (Nicotinamide adenine denucleotide (NADH) oxidase, all of which are involved in ROS generation [68, 15].

A number of flavonoids efficiently chelate trace metals, which play an important role in oxygen metabolism. Free iron and copper are potential enhancers of ROS formation, as exemplified by the reduction of hydrogen peroxide with the generation of the highly aggressive hydroxyl radical [103].

On the other hand, *in* vitro studies showed that the compounds present in fruits and vegetables, such as resveratrol, genestein, baicalein, and many others are attractive candidates for improved chemotherapeutic agents [35]. Resveratrol in combination with platinum drugs and oxaliplatin demonstrated that resveratrol administered 2 h prior to platinum drugs may sensitize ovarian cancer cells to platinum, inducing apoptosis and providing a means of overcoming resistance [95].

Ren [109] demonstrated that (–)–epigallocatechin-3-gallate induces reduction in IM9 myeloma cells and that its activity was dose- and time-dependent on the induction of apoptotic cell death; additionally, this natural metabolite combined with curcumin and lovastatin possessed the ability to suppress esophageal cancer-cell growth [154]. In multilla berries, it was found that their high levels of polyphenols, flavonoids, and flavonols and their antioxidants have a strong ability to reduce the viability of colon-cancer HT-29 and SW480 cell lines [33]. The anticancer activity of baicalein, a flavonoid found in several plants, was evaluated in a cutaneous squamous carcinoma-cell line, A431; it was found that this compound reduced the migration and invasiveness of the cells through inhibition of ezrin expression, which leads to the suppression of tumor metastasis [153].

In beans, it was found that these contain several compounds with cytotoxic activity on animals and human cell lines (C33-A, SW480, and 3T3), which can be attributed to the antioxidants and damage to DNA caused by tannins, saponins, lectins, and others compounds found in the seed [143, 144].

Melastoma malabathricum showed to have the ability to inhibit the proliferation of Caov-3, HL-60, CEM-SS, MCF-7, HeLa, and MDA-MB-231 cell lines, indicating that the leaves of this plant possess potential antiproliferative and antioxidant activities that could be attributed to its high content of phenolic compounds [122]. Melatonin, a naturally occurring compound, showed cytotoxic activity toward transformed 3T3-SV40 fibroblasts [143] and murine hepatoma cells MN22a, and it was shown that the sensitivities of both cell types to lysis by killer cells fell sharply [139].T he potent antioxidant activity of *Kalanchoe gracilis* (L.) DC stems due to that the polyphenolic compound found in this medicinal plant showed to have the ability to inhibit HepG2 cell proliferation [171], and the flavonoids found in *Rosa canina* L. are responsible for the antiproliferative activity in HeLa, MCF7, and HT-29 cancer-cell lines [138]. Analysis of the fruit of *Phelaria macrocarpa* (Boerl.) Scheff and of *Olea europaea* L. indicated that all parts of the fruit possess cytotoxic activity against HT-29, MCF-7, HeLa, BPH-1, and Chang cells, indicating that these fruits are a sources of bioactive compounds that are as potent as antioxidants and antioxidant agents, suggesting its possible use as an adjuvant agent in the treatment of cancer [56, 1].

The extract of *Calluna vulgaris* exhibited a photoprotective effect on human keratinocytes (HaCaT) exposed to Ultraviolet B (UVB) radiation [100]. *Cachrys pungens Jan* was analyzed in a human tumor- cell line, amelanotic melanoma, and it was found that its extract contains antioxidants, such as coumarins, which are responsible for their cytotoxicity in A375 cells [92]. *Inonotus obliquus* and *Peperomia pellucida*, plants employed as folk remedies for cancer treatment, were evaluated in several tumor cell-line types and it was found that these plants contains several antioxidants, such as lanosterol, inotodiols, ergosterol, phytol, 2-naphthalenol, decahydro hexadecanoic acid, methyl ester, and 9,12 octadecadienoic acid, indicating that these antioxidant compounds are responsible for the anticarcinogenic activity of the plant extract [129, 150]. The extract of *Indigofera cassioides* indicated the presence antioxidant activity, preventing lipid peroxidation and promoting the enzymatic antioxidant defense system, and also showed potent antitumoral and cytotoxic affect against EAC, DLA, HeLa, Hep-2, HepG-2, MCF-7, Ht-29, and NIH 3T3 cells [69].

Hesperetin, hesperetin analog, carnocine, and resveratrol were evaluated for their antioxidant and anticarcinogenic activity on HT-29, HCT116, and mouse skin carcinogenesis; their studies demonstrated that these compounds can inhibit cell proliferation, induce apoptosis, affect glycolysis, and decrease tumoration [125, 161, 40]. Honey, a natural product commonly used throughout the world, contains antioxidant properties and exerts a preventive effect against disease. Chrysin is a natural flavone commonly found in honey, and it was demonstrated that this compound induced apoptosis in PC-3 cells [116], fennel seeds (*Foeniculum vulgare*) are present in antioxidants that have an anticancer potential against HepG2 and MCF-7 cell lines [94]. It was indicated that compounds such as quercetin, flavonoids, and brucine have chemopreventive action against the osteosarcoma cell line (MG63), C6 glioma cells, and Ehrlich ascites cells, and that they can be used as anticancer, antigenotoxic agents and can induce apoptosis [135, 119, 2].

8. Conclusion

Oxidative stress causes injury to cells, induces gene mutation, and is involved in carcinogenesis and other degenerative diseases by directly or indirectly influencing intracellular signal transduction and transcription factors. The state of OS under carcinogenesis and tumor-bearing conditions is an intricate one in which various substances are involved in complex interactions.

The data discussed in this paper show that the biological effects of antioxidants on humans and animals can be controversial. Due to that the action of antioxidants depends on the oxidative status of cells, antioxidants can be protective against cancer; because ROS induce oxidative carcinogenic damage in DNA, antioxidants can prevent cancer in healthy persons harboring increased ROS levels.

Oxidative stress as cause and effect is not the sole factor in the development of cancer. It is important to take into account that there are other factors involved in its development, such as genetic predisposition, eating habits, environment, etc. Because ROS at moderate concentrations act as indispensable mediators of cancer-protective apoptosis and phagocytosis, an excess of antioxidants in persons with low ROS levels can block these cancer-preventive mechanisms. High doses of antioxidants can reduce the ROS level in persons who overproduce ROS and protect them against cancer and other ROS-dependent morbid conditions.

For individuals with low ROS levels, high doses of antioxidants can be deleterious, suppressing the already low rate of ROS generation and ROS-dependent cancer-preventive apoptosis. Screening and monitoring the human population regarding their ROS level can transform antioxidants into safe and powerful disease-preventive tools that could significantly contribute to the nation's health.

Many *in vivo* and *in vitro* studies performed to evaluate the capability of antioxidants against cancer, such as chemopreventive or therapeutic agents, were conduced employing natural antioxidants from fruits and vegetables; these are mainly supplied through food, which often do not provide sufficient input for these to function as chemoprotectors. Thus, humans are forced to consume antioxidants in a more direct manner, either in the form of a tablet, a pill, or any other form in order to supply the levels that the body requires of these compounds to protect it against cell damage caused by oxidation reactions, thus reducing the risk of certain cancer types, especially those of the epithelial surface and in the upper part of the body, such as breast, lung, kidney, liver, intestine, and many others that have been well documented. However, further investigations are expected before our better understanding of the function of many antioxidants and their utilization in the prevention and treatment of cancer and other degenerative diseases.

Author details

Carmen Valadez-Vega[1], Luis Delgado-Olivares[1], José A. Morales González[1], Ernesto Alanís García[1], José Roberto Villagomez Ibarra[2], Esther Ramírez Moreno [1], Manuel Sánchez Gutiérrez[1], María Teresa Sumaya Martínez[3], Zuñiga Pérez Clara[1] and Zuli Calderón Ramos[1]

1 Institute of Health Sciences, Autonomous University of Hidalgo State, Ex-Hacienda de la Concepción, Tilcuautla, Hgo, Mexico. C.P.42080., Mexico

2 Institute of Basic Sciences, Autonomous University of Hidalgo State, Km 4.5 Carretera Pachuca-Tulancingo, Ciudad del Conocimiento, Mineral de la Reforma Hidalgo, C.P. 42076, Mexico

3 Secretary of Research and Graduate Studies, Autonomous University of Nayarit, Ciudad de la Cultura "Amado Nervo", Boulevard Tepic-Xalisco S/N. Tepic, Nayarit, Mexico

References

[1] Acquaviva, R, Di Giacomo, C, Sorrenti, V, Galvano, F, Santangelo, R, Cardile, V, Gangia, S, D'Orazio, N, Abraham, NG, & Vanella, L. (2012). Antiproliferative effect of oleuropein in prostate cell lines. *International Journal of Oncology*, Print, 1791-2423, Online, 1019-6439, 41, 31-38.

[2] Agrawal, S. S., Saraswati, S., Mathur, R., & Pandey, M. (2011). Cytotoxic and antitumor effects of brucine on Ehrlich ascites tumor and human cancer cell line. *Life Science*, 89, 147-158, 0024-3205.

[3] Akah, P. A., & Ekekwe, R. K. (1995). Ethnopharmacology of some of the asteraceae family used in the Nigerian tradition al medicine. *Fitoterapia*, 66, 352-355, 0036-7326 X.

[4] Akinpelu, D. A. (1999). Antimicrobial activity of Vernonia amygdalina leaves. *Fitoterapia*, 70, 232-234, 0036-7326 X.

[5] Allen, R. G., & Tresini, M. (2000). Oxidative stress and gene regulation. *Free Radical Biology & Medicine*, 28, 463-499, 0891-5849.

[6] Ando, K., & Fujita, T. (2009). Metabolic Syndrome and Oxidative Stress. *Free Radical Biology & Medicine*, 47, 213-218, 0891-5849.

[7] Andreazza, AC, Kapczinski, F, Kauer-Sant'Anna, M, Walz, JC, Bond, DJ, Gonçalves, CA, Young, LT, & Yatham, LN. (2009). 3-Nitrotyrosine and glutathione antioxidant system in patients in the early and late stages of bipolar disorder. *Journal of Psychiatry and Neuroscience*, 1488-2434, 4, 263-271.

[8] Anusuya, C, & Manoharan, S. (2011). Antitumor initiating potential of rosmarinic acid in 7,12-dimethylbenz(a)anthracene-induced hamster buccal pouch carcinogenesis. *Journal of Environmental Pathology, Toxicology and Oncology*, Print, 0731-8898, Online, 2162-6537, 30, 199-211.

[9] Asokkumar, S., Naveenkumar, C., Raghunandhakumar, S., Kamaraj, S., Anandakumar, P., Jagan, S., & Devaki, T. (2012). Antiproliferative and antioxidant potential of beta-ionone against benzo(a)pyrene-induced lung carcinogenesis in Swiss albino mice. *Molecular and Cellular Biochemistry*, 363, 335-345, 0300-8177, Print, 1573-4919, (Online).

[10] Bartoli, R., Fernandez-Banares, F., Navarro, E., Castella, E., Mane, J., Alvarez, M., Pastor, C., Cabre, E., & Gassull, M.A. (2000). Effect of olive oil on early and late events of colon carcinogenesis in rats: Modulation of arachidonic acid metabolism and local prostaglandin E(2) synthesis. *Gut*, 46, 191-199, 0017-5749, Print, 1468-3288, (Online).

[11] Berlett, BS, & Stadtman, E. R. (1997). Protein Oxidation in Aging, Disease, and Oxidative Stress. *The Journal Of Biological Chemistry*, 272(33), 20313-20316, 0021-9258, Print, 1083-351X, (Online).

[12] Bhagat, S. S., Ghone, R. A., Suryakar, A. N., & Hundekar, P. S. (2011). Lipid peroxidation and antioxidant vitamin status in colorectal cancer patients. *Indian Journal Physiology and Pharmacology*, 55, 72-76, 0019-5499.

[13] Bohr, V., Anson, S., Mazur, R. M., & Dianov, G. (1998). Oxidative DNA damage processing and changes with aging. *Toxicology Letters Vols*, 102-103, 47-52, 0378-4274.

[14] Boon, H., Stewart, M., Kennard, MA, Gray, R., Sawka, C., Brown, J. B., Mc William, C., Garvin, A., Baron, R. A., Aaron, D., & Haines-Kamka, T. (2000). Use of complementary/alternative medicine by breast cancer survivors in Ontario: prevalence and perceptions. *Journal of Clinical Oncology*, 8, 2515-2521, 0073-2183 X.

[15] Brown, J. E., Khodr, H., Hider, R. C., & Rice-Evans, C. (1998). Structural dependence of flavonoid interactions with Cu2+ ions: implications for their antioxidant properties. *Biochemical Journal*, 330, 1173-1178, 0264-6021, Print, 1470-8728, (Online).

[16] Brown, N. S., & Bicknell, R. (2001). Hypoxia and oxidative stress in breast cancer: Oxidative stress: its effects on the growth, metastatic potential and response to therapy of breast cancer. *Breast Cancer Research*, 3, 323-327, 0167-6806, Print, 1573-7217, (Online).

[17] Cadet, J., Douki, T., & Ravanat, J. L. (1997). Artifacts associated with the measurement of oxidized DNA bases. *Environmental Health Perspectives*, 105, 1034-1039, 0091-6765.

[18] Calomme, M., Pieters, L., Vlietinck, A., & Vanden, Berghe. D. (1996). Inhibition of bacterial mutagenesis by Citrus flavonoids. *Planta Medica*, 62, 222-226, 0032-0943.

[19] Carlsen, M. H., Halvorsen, B. L., Holte, K., Bøhn, S. K., Dragland, S., Sampson, L., Willey, C., Senoo, H., Umezono, Y., Sanada, C., Barikmo, I., Berhe, N., Willett, W. C., Phillips, K., Jacobs, D. R. Jr, & Blomhoff, R. (2010). The total antioxidant content of more than 3100 foods, beverages, spices, herbs and supplements used worldwide. *Nutrition Journal*, 9(3), http://www.nutritionj.com/content/9/1/3, 1475-2891.

[20] Cejas, P., Casado, E., Belda-Iniesta, C., De Castro, J., Espinosa, E., Redondo, A., Sereno, M., García-Cabezas, M. A., Vara, J. A., Domínguez-Cáceres, A., Perona, R., & González-Barón, M. (2004). Implications of oxidative stress and cell membrane lipid peroxidation in human cancer (Spain). *Cancer Causes and Control*, 15, 707-719, 0957-5243, Print, 1573-7225, (Online).

[21] Chen, L, Hu, JY, & Wang, SQ. (2012). The role of antioxidants in photoprotection: A critical review. *Journal of the American Academy of Dermatology*, 10.1016/j.jaad. 2012.02.009, [Epub ahead of print], 0190-9622, 0190-9622.

[22] Chinery, R., Brockman, J. A., Peeler, M. O., Shyr, Y., Beauchamp, R. D., & Coffey, R. J. (1997). Antioxidants enhance the cytotoxicity of chemotherapeutic agents in colorectal cancer: a p53-independent induction of p21WAF1/CIP1 via C/EBP. *Nature Medicine*, 3, 1233-1241, 1078-8956.

[23] Dal-Pizzol, F., Ritter, C., Cassol Jr, Oj., Rezin, G. T., Petronilho, F., Zugno, A. I., Que-vedo, J., & Streck, E. L. (2009). Oxidative Mechanisms of Brain Dysfunction During Sepsis. *Neurochemical Research*, 35, 1-12, DOI: s11064-009-0043-4, 0364-3190, Print, 1573-6903, (Online).

[24] Delgado, O. L., Betanzos, C. G., & Sumaya, M. M. T. (2010). Importancia de los anti-oxidantes dietarios en la disminución del estrés oxidativo. *Investigación y Ciencia*, 50, 10-15, 1665-4412.

[25] De Mejia, E. G., Valadez-Vega, M. C., Reynoso-Camacho, R., & Loarca-Pina, G. (2005). Tannins, trypsin inhibitors and lectin cytotoxicity in tepary (Phaseolus acuti-folius) and common (Phaseolus vulgaris) beans. *Plant Foods Hum Nutr*, 60, 137-145, 0921-9668.

[26] Di Domenico, F, Foppoli, Coccia, C, R, & Perluigi, M. (2012). Antioxidants in cervical cancer: Chemopreventive and chemotherapeutic effects of polyphenols. *Biochimica et Biophysica Acta*, 0005-2736, 1822, 737-747.

[27] Drisko, J. A., Chapman, J., & Hunter, V. J. (2003). The use of antioxidants with first-line chemotherapy in two cases of ovarian cancer. *Journal of the American College of Nutrition*, 22, 118-123, 1665-4412.

[28] Duthie, S. J., & Dobson, V. L. (1999). Dietary flavonoids protect human colonocyte DNA from oxidative attack in vitro. *European Journal of Nutrition*, 38, 28-34, 0022-3166, Print, 1541-6100, (Online).

[29] Ebenezer, O., Farombi, A., & Olatunde. (2011). Antioxidative and chemopreventive properties of Vernonia amygdalina and Garcinia biflavonoid. *International Juornal of Environment Researc and Public Health*, 8, 2533-2555, 1661-7827, Print, 1660-4601, (On-line).

[30] Echart, M. A. M., Barrio, L. J. P., Maria, Gabriela., Valle, G. M. G., Augustin, S. C. H., Ugalde Marques da Rocha, MI, Manica-Cattani, MF, Feyl dos Santos, G, & Manica da Cruz, IB. (2009). Association between manganese superoxide dismutase (MnSOD). gene polymorphism and elderly obesity. *Molecular and Cellular Biochemistry*, 328, 33-40, 0300-8177, Print, 1573-4919, (Online).

[31] Ernst, E., & Cassileth, B. R. (1998). The prevalence of complementary/alternative medicine in cancer: a systematic review. *Cancer*, 83, 777-782, 0000-8543 X, (Print), 1097-0142, (Online).

[32] Esterbauer, H., Schauur, J. S., & Zollner, H. (1991). Chemistry and biochemistry of 4-hydroxynonenal, malonaldehyde and related aldehydes. *Free Radical Biology & Medi-cine*, 11, 81-128, 0891-5849.

[33] Flis, S., Jastrzebski, Z., Namiesnik, J., Arancibia-Avila, P., Toledo, F., Leontowicz, H., Leontowicz, M., Suhaj, M., Trakhtenberg, S., & Gorinstein, S. (2012). Evaluation of in-hibition of cancer cell proliferation in vitro with different berries and correlation with

their antioxidant levels by advanced analytical methods. *Journal of Pharmaceutical Biomedical Analysis*, 62, 68-78, 0731-7085.

[34] Floyd, R. A., Watson, J. J., & Wong, P. K. (1986). Hydroxyl free radical adduct of deoxyguanosine: sensitive detection and mechanisms of formation. *Free Radical Research Communications*, 1, 163-172, 8755-0199.

[35] Fox, J. T., Sakamuru, S., Huang, R., Teneva, N., Simmons, S. O., Xia, M., Tice, R. R., Austin, , & Myung, K. (2012). High-throughput genotoxicity assay identifies antioxidants as inducers of DNA damage response and cell death. *Proceedings of the National Academy of Sciences of United States of America*, 109, 5423-5428.

[36] Fuchs-Tarlovsky, V., Bejarano-Rosales, M., Gutierrez-Salmeán, G., Casillas, MA, López-Alvarenga, J. C., & Ceballos-Reyes, G. M. (2011). Effect of antioxidant supplementation over oxidative stress and quality of life in cervical cancer. *Nutrición Hospitalaria*, 26, 819-826, 0212-1611.

[37] Fukumura, H, Sato, M, Kezuka, K, Sato, I, Feng, X, Okumura, S, Fujita, T, Yokoyama, U, Eguchi, H, Ishikawa, Y, & Saito, T. (2012). Effect of ascorbic acid on reactive oxygen species production in chemotherapy and hyperthermia in prostate cancer cells. *The Jornal of Physiological Sciences*, 1880-6546, (Print), 1880-6562, Online, 62, 251-257.

[38] Gago-Dominguez, M., Jiang, X., & Castelao, J. E. (2007). Lipid peroxidation, oxidative stress genes and dietary factors in breast cancer protection: a hypothesis. *Breast Cancer*, 9, 1-11, 10.1186/bcr1628, http://breast-cancer-research.com/content/9/1/201, 0146-5542 X.

[39] Gamaleǐ, I. A., Kirpichnikova, K. M., & Filatova, N. A. (2011). Effect of melatonin on the functional properties of transformed cells. *Vopr Onkol*, 57, 481-485, 0507-3758.

[40] George, J, Singh, M, Srivastava, AK, Bhui, K, Roy, P, Chaturvedi, PK, & Shukla, Y. (2011). Resveratrol and black tea polyphenol combination synergistically suppress mouse skin tumors growth by inhibition of activated MAPKs and p53. *PLoS One*, 1932-6203, 6, 23395-23408.

[41] Gibellini, L., Pinti, M., Nasi, M., Montagna, J. P., De Biasi, S., Roat, E., Bertoncelli, L., Cooper, E. L., & Cossarizza, A. (2011). Quercetin and cancer chemoprevention. *Evidence-Based Complementary and Alternative Medicine*, 59, 1356-1365, 0174-1427, Print, 1741-4588, (Online).

[42] Gill, C. I., Boyd, A., Mc Dermott, E., Mc Cann, M., Servili, M., Selvaggini, R., Taticchi, A., Esposto, S., Montedoro, G., Mc Glynn, H., & Rowland, I. (2005). Potential anticancer effects of virgin olive oil phenols on colorectal carcinogenesis models in vitro. *International Journal of Cancer*, 117, 1-7, 0020-7136, 1097-0215, (Online).

[43] Gille, G, & Sigler, K. (1995). Oxidative stress and living cells. *Folia Microbiological*, 0015-5632, (Print), 1874-9356, (Online), 40, 131-152.

[44] Gong, G., Waris, G., Tanveer, R., & Siddiqui, A. (2001). Human hepatitis C virus NS5A protein alters intracellular calcium levels, induces oxidative stress, and acti-

vates STAT-3 and NF-B. *Proceedings of the National Academy of Sciences of United States of America,* 98(17), 9599-9604, 0027-8424.

[45] Grimsrud, P. A., Xie, H., Griffin, T. J., & Bernlohr, D. A. (2008). Oxidative Stress and Covalent Modification of Protein with Bioactive Aldehydes. *Journal of Biological Chemistry,* 283(32), 21837-21841, 0021-9258, (Print), 1083-351X, (Online).

[46] Gupta, V, & Sharma, M. (2012). Phytochemical Analysis and Evaluation of Antioxidant Activities of Methanolic Extracts of Maytenus emarginata. 1536-2310, (Print), 1557-8100, Online, 16(5), 257-262.

[47] Halliwell, B. (1994). Free radicals, antioxidants, and human disease: curiosity, cause, or consequence? *The Lancet,* 344, 721-724, 1040-6736.

[48] Halliwell, B. (1996). Antioxidants in Human Health and Disease. Annual Reviews, 1550-8382 Online , 16, 33-50.

[49] Halliwell, B. (2000). The antioxidant paradox. *The Lancet,* 1, 1179-1180, 1040-6736.

[50] Halliwell, B., & Gutteridge, J. M. C. (2006). Free Radicals in Biology and Medicine. Ed 4. Clarendon Press, Oxford.

[51] Halliwell, B. (2007). Oxidative stress and cancer: have we moved forward? *Biochemical Journal,* 401, 1-11, 0264-6021, Print, 1470-8728, (Online).

[52] Halliwell, B. (2008). Are polyphenols antioxidants or pro-oxidants? What do we learn from cell culture and in vivo studies? *Archives of Biochemistry and Biophysics,* 476, 107-112, 0003-9861, Print, 1096-0384, (Online).

[53] Han, R.M., Tian, Y.X., Becker, E.M., Andersen, M.L., Zhang, J.P., & Skibsted, L.H. (2007). Puerarin and conjugate bases as radical scavengers and antioxidants: molecular mechanism and synergism with beta-carotene. *Journal of Agricultural and Food Chemistry,* 0021-8561, Print, 1520-5118, Online, 55, 2384-2389.

[54] Hanasaki, Y., Ogawa, S., & Fukui, S. (1994). The correlation between active oxygens scavenging and antioxidative effects of flavonoids. *Free Radical Biology & Medicine,* 16, 845-850, 0891-5849.

[55] Hashim, Y. Z., Rowland, I. R., Mc Glynn, H., Servili, M., Selvaggini, R., Taticchi, A., Esposto, S., Montedoro, G., Kaisalo, L., Wahala, K., & Gill, C. I. (2008). Inhibitory effects of olive oil phenolics on invasion in human colon adenocarcinoma cells in vitro. *International Journal of Cancer,* 122, 495-500, 0020-7136, Print, 10970215, Online.

[56] Hendra, R., Ahmad, S., Oskoueian, E., Sukari, A., & Shukor, M. Y. (2011). Antioxidant, anti-inflammatory and cytotoxicity of Phaleria macrocarpa (Boerl.) Scheff Fruit. *BMC Complemententary & Alternative Medicine,* 11, 110-121, 1472-6882.

[57] Hippeli, S., Heiser, I., & Elstner, E. F. (1999). Activated oxygen and free oxygen radicals in pathology: New insights and analogies between animals and plants. *Plant Physiology Biochemistry,* 37, 167-178, 0981-9428.

[58] Hladik, C., Krief, S., & Haxaire, C. (2005). Ethnomedicinal and bioactive properties of plants ingested by wild chimpanzees in Uganda. *Journal Ethnopharmacology*, 101, 1-5, 0378-8741.

[59] Hu, M.L. (2011). Dietary Polyphenols as Antioxidants and Anticancer Agents: More Questions than Answers. *Chang Gung Medical Journa*, 2072-0939, 34, 449-459.

[60] Inoue, M., Tajima, K., Mizutani, M., Iwata, H., Iwase, T., Miura, S., Hirose, K., Hamajima, N., & Tominaga, S. (2001). Regular consumption of green tea and the risk of breast cancer recurrence: Follow-up study from the Hospital-based Epidemiologic Research Program at Aichi Cancer Center (HERPACC), Japan. *Cancer Letters*, 167, 175-182, 0304-3835.

[61] Iovine, B., Iannella, M. L., Nocella, F., Pricolo, M. R., & Bevilacqua, MA. (2012). Carnosine inhibits KRAS-mediated HCT116 proliferation by affecting ATP and ROS production. *Cancer Letters*, 28, 122-128, 0304-3835.

[62] Jayaprakash, V., & Marshall, J. R. (2011). Selenium and other antioxidants for chemoprevention of gastrointestinal cancers. *Best Practtice & Research Clinical Gastroenterology*, 25, 507-518, 1521-6918.

[63] Jisaka, M., Ohigashi, H., Takegawa, K., Hirota, M., Irie, R., Huffman, MA, & Koshimizu, K. (1993). Steroid glucosides from Vernonia amygdalina, a possible chimpanzee plant. *Phytochemistry*, 34, 409-413, 0031-9422.

[64] Katiyar, S. K., & Mukhtar, H. (1997). Tea antioxidants in cancer chemoprevention. *Journal of Cellular Biochemistry*, 27, 59-67, 1097-4644.

[65] Keller, J.N. (2006). Interplay Between Oxidative Damage, Protein Synthesis, and Protein Degradation in Alzheimer's Disease. *Journal of Biomedicine and Biotechnology*, http://www.ncbi.nlm.nih.gov/pmc/articles/PMC1510934/pdf/JBB2006-12129.pdf, 1110-7243, Print, 1110-7251, Online, 2006, 1-3.

[66] Khan, N., Afaq, F., Saleem, M., Ahmad, N., & Mukhtar, H. (2006). Targeting multiple signaling pathways by green tea polyphenol (−)-epigallocatechin-3-gallate. *Cancer Research*, 66, 2500-2505, 0008-5472, Print, 1538-7445, Online.

[67] Kleinschnitz, C., Grund, H., Wingler, K., Armitage, ME, Jones, J., Mittal, M., Barit, D., Schwarz, T., Geis, C., Kraft, P., Barthel, K., Schuhmann, M. K., Herrmann, A. M., Meuth, S. G., Stoll, G., Meurer, S., Schrewe, A., Becker, L., Gailus-Durner, V., Fuchs, H., Klopstock, T., Hrabe′ de Angelis, M., Jandeleit-Dahm, K., Shah, A. M., Weissmann, N., & Schmidt, H. H. H. W. (2010). Post-Stroke Inhibition of Induced NADPH Oxidase Type 4Prevents Oxidative Stress and Neurodegeneration. *PloS Biology*, 8(9), http://www.plosbiology.org/article/info%3Adoi%2F10.1371%2Fjournal.pbio.1000479, 1545-7885, 1544-9173.

[68] Korkina, L. G., & Afanas′ev, I. B. (1997). *Antioxidants in Disease Mechanisms and Therapy*, Sies, H., Ed.; Academic Press: San Diego, 151-163.

[69] Kumar, RS, Rajkapoor, B, & Perumal, P. (2011). In vitro and in vivo anticancer activity of Indigofera cassioides Rottl. *Ex. DC. Asian Pacific Journal of Tropical Medicine*,
1995-7645, 4, 379-385.

[70] La Vecchia, C., Altieri, A., & Tavani, A. (2001). Vegetables, fruit, antioxidants and
cancer: a review of Italian studies. *European Journal of Nutrition*, 40, 261-267,
1436-6207, Print, 1436-6215, Online.

[71] Lai, Z. R., Ho, Y. L., Huang, S. C., Huang, T. H., Lai, S. C., Tsai, J. C., Wang, C. Y.,
Huang, G. J., & Chang, Y. S. (2011). Antioxidant, anti-inflammatory and antiprolifer-
ative activities of Kalanchoe gracilis (L.) DC stem. *The American Journal of Chinese
Medicine*, 39, 1275-1290, 0019-2415 X, Print, 1793-6853, Online.

[72] Larsen, CA, & Dashwood, R. H. (2009). Suppression of Met activation in human colon cancer cells treated with (−)-epigallocatechin-3-gallate: Minor role of hydrogen
peroxide. *Biochemical and Biophysical Research Communnications*, 389, 527-530,
0000-6291 X.

[73] Lee, C. K., Weindruch, R., & Prolla, T. A. (2000). Gene-expression profile of the ageing brain in mice. *Nature Genetics*, 25, 294-297, 1061-4036.

[74] Leong, H., Mathur, P. S., & Greene, G. L. (2008). Inhibition of mammary tumorigene-
sis in the C3(1)/SV40 mouse model by green tea. *Breast Cancer Research and Treatment*,
107, 359-369, 0167-6806, Print, 0167-6806, Print, 1573-7217, Online.

[75] Li, Q., Zhao, H. F., Zhang, Z. F., Liu, Z. G., Pei, X. R., Wang, J. B., Cai, M. Y., & Li, Y.
(2009). Long-term administration of green tea catechins prevents age-related spatial
learning and memory decline in C57BL/6 J mice by regulating hippocampal cyclic
AMP-response element binding protein signaling cascade. *Neuroscience*, 159,
1208-1215, 0306-4522.

[76] Li, W., Shi, Y. H., Yang, R. L., Cui, J., Xiao, Y., Wang, B., & Le , G. W. (2010). Effect of
somatostatin analog on high-fat diet-induced metabolic syndrome: Involvement of
reactive oxygen species. *Peptides*, 31(4), 625-629, 0196-9781.

[77] Liang, W., Li, X., Li, C., Liao, L., Gao, B., Gan, H., Yang, Z., Liao, L., & Chen, X.
(2011). Quercetin-mediated apoptosis via activation of the mitochondrial-dependent
pathway in MG-63 osteosarcoma cells. *Molecular Medicine Reports*, 4, 1017-1023,
1791-2997, Print, 1791-3004, Online.

[78] Liu, M., Gong, X., Alluri, R. K., Wu, J., Sablo, T., & Li, Z. (2012). Characterization of
RNA damage under oxidative stress in Escherichia coli. *Biol Chem*, 393(3), 123-132,
1437-4315.

[79] Llor, X., Pons, E., Roca, A., Alvarez, M., Mane, J., Fernandez-Banares, F., & Gassull,
M. A. (2003). The effects of fish oil, olive oil, oleic acid and linoleic acid on colorectal
neoplastic processes. *Clinical Nutrition*, 22, 71-79, 0261-5614.

[80] Lockwood, K., Moesgaard, S., Hanioka, T., & Folkers, K. (1994). Apparent partial remission of breast cancer in 'High Risk' patients supplemented with nutritional anti-

oxidants, essential fatty acids and Coenzyme Q_{10} . *Biochemical and Biophysical Research Communications*, 15, 231-s240, 0000-6291 X.

[81] Ma, Q, & Kinneer, K. (2002). Chemoprotection by phenolic antioxidants. Inhibition of tumor necrosis factor alpha induction in macrophages. *Journal of Biological Chemistry*, 0021-9258, Print, 1083-351X, Online, 277, 2477-2484.

[82] Maritim, A. C., Sanders, R. A., & Watkins, I. I. I. J. B. (2003). Diabetes, Oxidative Stress, and Antioxidants: A Review. *Journal of Biochememical Molecular and Toxicology*, 17, 24-38, 1095-6670, Print, 1099-0461, Online.

[83] Markesbery, W. R. (1997). Oxidative Stress Hypothesis In Alzheimer's Disease. *Free Radical Biology & Medicine*, 23(1), 134-147, 0891-5849.

[84] Martinez, M. E. (2005). Primary prevention of colorectal cancer: Lifestyle, nutrition, exercise. *Recent Results in Cancer Research*, 166, 177-211, 0080-0015.

[85] Matés, JM, Segura, JA, Alonso, FJ, & Márquez, J. (2011). Anticancer antioxidant regulatory functions of phytochemicals. *Current Medicinal Chemistry*, 0929-8673, Print, 1875-533X, Online, 18, 2315-2338.

[86] Mates, J. M., Perez-Gomez, C., & Nunez de Castro, I. (1999). Antioxidant enzymes and human diseases. *Clinical Biochemistry*, 32, 595-603, 0009-9120.

[87] Mauro, S., Rino, B., Alicja, W., & Anna, (2002. (2002). Total antioxidant potential of fruit and vegetables and risk of gastric cancer. *Gastroenterology*, 123, 985-991, 0016-5085.

[88] Maxmen, A. (2012). The Hard Facts. *Nature*, 485, S50-S51, 0028-0836.

[89] Mazdak, H., & Zia, H. (2012). Vitamin e reduces superficial bladder cancer recurrence: a randomized controlled trial. *International Journal of Preventive Medicine*, 3, 110-115.

[90] McCord, J. M., & Fridovich, I. (1968). The reduction of cytochrome c by milk xanthine oxidase. *The Journal of Bioogical Chemistry*, 2008-7802, Print, 2008-8213, Online, 243, 5753-5760.

[91] Medina-Ceja, L., Guerrero-Cazares, H., Canales-Aguirre, A., Morales-Villagrán, A., & Feria-Velasco, A. (2007). Características estructurales y funcionales de los transportadores de glutamato: su relación con la epilepsia y el estrés oxidativo. *Revista de Neurología*, 45(6), 341-352.

[92] Menichini, G, Alfano, C, Provenzano, E, Marrelli, M, Statti, GA, Menichini, F, & Conforti, F. (2012). Cachrys pungens Jan inhibits human melanoma cell proliferation through photo-induced cytotoxic activity. *Cell Proliferation*, 0960-7722, Print, 1365-2184, Online, 45, 39-47.

[93] Middleton, E. Jr, Kandaswami, C., & Theoharides, T. C. (2000). The effects of plant flavonoids on mammalian cells: Implications for inflammation, heart disease, and cancer. *Pharmacological Reviews*, 52, 673-751, 0031-6997.

[94] Mohamad, R. H., El -Bastawesy, A. M., Abdel-Monem, M. G., Noor, A. M., Al-Mehdar, H. A., Sharawy, S. M., & El -Merzabani, MM. (2011). Antioxidant and anticarcinogenic effects of methanolic extract and volatile oil of fennel seeds (Foeniculum vulgare). *Journal of Medicine Food,*, 14, 986-1001, 0109-6620, Print, 1557-7600, Online.

[95] Nessa, M. U., Beale, P., Chan, C., Yum, J. Q., & Huq, F. (2012). Combinations of resveratrol, cisplatin and oxaliplatin applied to human ovarian cancer cells. *Anticancer Res*, 32, 53-59, 0250-7005, Print, 1791-7530, Online.

[96] Noda, N., & Wakasugi, H. (2000). Cancer and oxidative stress. *Journal of the Japan Medical Association*, 124(11), 1571-1574, 1356-8650.

[97] Nunomura, A., Honda, K., Takeda, A., Hirai, K., Zhu, X., Smith, M. A., & Perry, G. (2006). Oxidative Damage to RNA in Neurodegenerative Diseases. *Journal of Biomedicine and Biotechnology* [82323], 1-6, 1110-7243, Print, 1110-7251, Online.

[98] Nyström, N. (2005). Role of oxidative carbonylation in protein quality control and senescence. *EMBO Journal*, 0261-4189, Print, 1460-2075, Online, 24, 1311-1317.

[99] Owen, R. W., Giacosa, A., Hull, W. E., Haubner, R., Spiegelhalder, B., & Bartsch, H. (2000). The antioxidant/anticancer potential of phenolic compounds isolated from olive oil. *European Journal Cancer*, 36, 1235-1247, 0959-8049.

[100] Perde-Schrepler, M, Chereches, G, Brie, L, Virag, P, Barbo,s, O, Soritau, O, Tatomir, C, Fischer-Fodor, E, Filip, A, Vlase, L, & Postescu, ID. (2011). Photoprotective effect of Calluna vulgaris extract against UVB-induced phototoxicity in human immortalized keratinocytes. *Journal of Environment Pathology Toxicology and Oncology*, 0371-8898, Print, 2162-6537, Online, 30, 323-331.

[101] Perera, R. M., & Bardeesy, N. (2011). When antioxidants are bad. *Nature*, 4, 4, 0028-0836.

[102] Pietá, D., Martins De, Lima. M. N., Presti-Torres, J., Dornelles, A., Garcia, V. A., Siciliani, S. F., Rewsaat, M. G., Constantino, L., Budni, P., Dal-Pizzol, F., & Schrödera, N. (2007). Memantine Reduces Oxidative Damage And Enhances Long-Term Recognition Memory In Aged Rats. *Neuroscience*, 146, 1719-1725, 0306-4522.

[103] Pietta, P. G. (2000). Flavonoids as Antioxidants. *Journal of Natural Produts*, 1035-1042, 0163-3864, Print, 1520-6025, Online.

[104] Plaumann, B., Fritsche, M., Rimpler, H., Brandner, G., & Hess, R. D. (1996). Flavonoids activate wild-type p53. *Oncogene*, 13, 1605-1614, 0950-9232.

[105] Prasad, K. N., Kumarm, A., Kochupillaim, V., & Colem, W. C. (1999). High doses of multiple antioxidant vitamins: essential ingredients in improving the efficacy of standard cancer therapy. *Journal of the American College of Nutrition*, 18, 13-25, 0731-5724, Print, 1541-1087, Online.

[106] Prasad, K. N., Cole, W. C., Kumar, B., & Prasad, K. C. (2001). Scientific rationale for using high-dose multiple micronutrients as an adjunct to standard and experimental

cancer therapies. *Journal of the American College of Nutrition*, 20, 450S-463S, 0731-5724, Print, 1541-1087, Online.

[107] Price, T. O., Ercal, N., Nakaoke, R., & Banks, W. A. (2005). HIV-1viralproteins gp120 and Tatinduceoxidativestress in brain endothelial cells. *Brain Research*, 1045, 57-63, 0006-8993.

[108] Rabek, J. P., Boylston, I. I. I. W. H., & Papaconstantinou, J. (2003). Carbonylation of ER chaperone proteins in aged mouse liver. *Biochemical and Biophysical Research Communications*, 305, 566-572, 0000-6291 X.

[109] Ren, L., Yang, H. Y., Choi, H. I., Chung, K. J., Yang, U., Lee, I. K., Kim, H. J., Lee, , Park, B. J., & Lee, T. H. (2011). The role of peroxiredoxin V in (-)-epigallocatechin 3-gallate-induced multiple myeloma cell death. *Oncology Research*, 19, 391-398, 0965-0407.

[110] Rieger-Christ, KM, Hanley, R, Lodowsky, C, Bernier, T, Vemulapalli, P, Roth, M, Kim, J, Yee, AS, Le, SM, Marie, PJ, Libertino, JA, & Summerhayes, IC. (2007). The green tea compound, (-)-epigallocatechin-3-gallate downregulates N-cadherin and suppresses migration of bladder carcinoma cells. *Journal of Cellular Biochemistry*, 0730-2312, Print, 1097-4644, Online, 102, 377-388.

[111] Riordan, N. H., Riordan, H. D., Meng, Y. L., & Jackson, J. A. (1995). Intravenous ascorbate as a tumor cytotoxic. chemotherapeutic agent. *Medical Hypotheses*, 44, 207-213, 0306-9877.

[112] Riordan, N. H., Riordan, H. D., & Casciari, J. P. (2000). Clinical and experimental experiences with intravenous vitamin C. *Journal of Orthomolecular Medicine*, 5, 201-213, 0317-0219.

[113] Rivas, MA, Carnevale, R. P., Proietti, C. J., Rosemblit, C., Beguelin, W., Salatino, M., Charreau, E. H., Frahm, I., Sapia, S., Brouckaert, P., Elizalde, P. V., & Schillaci, R. (2008). TNF alpha acting on TNFR1 promotes breast cancer growth via P42 P44 MAPK, JNK, Akt and NF-kappa B-dependent pathways. *Experimental Cell Research*, 314(3), 509-29, 0014-4827.

[114] Roberts, C. K., Barnarda, R. J., Sindhub, R. K., Jurczak, M., Ehdaieb, A., & Vaziri, N. D. (2006). Oxidative stress and dysregulation of NAD(P)H oxidase and antioxidant enzymes in diet-induced metabolic syndrome. *Metabolism Clinical and Experimental*, 55, 928-934, 1532-8600.

[115] Roche, CE, & Romero, A. D. (1994). Estrés oxidativo y degradación. de proteínas. *Medicina clínica*, 103(5), 189-196, 0025-7753.

[116] Samarghandian, S, Afshari, JT, & Davoodi, S. (2011). Chrysin reduces proliferation and induces apoptosis in the human prostate cancer cell line pc-3. *Clinics (Sao Paulo)*, 1807-5932, Print, 1980-5322, Online, 66, 1073-1079.

[117] Schmitt, CA, & Lowe, S. W. (1999). Apoptosis and therapy. *The Journal of Pathololy*, 187, 127-137.

[118] Seef, L. B., Lindsay, K. L., Bacon, B. R., Kresina, F., & Hoofnagle, H. (2001). Comple-
       mentary and alternative medicine in chronic liver disease. *Hepatology*, 34, 595-603,
       1096-9896, Online.

[119] Seibert, H, Maser, E, Schweda, K, Seibert, S, & Gülden, M. (2011). Cytoprotective ac-
       tivity against peroxide-induced oxidative damage and cytotoxicity of flavonoids in
       C6 rat glioma cells. *Food and Chemical Toxicology*, 0278-6915, 49, 2398-2407.

[120] Sharma, R., & Vinayak, M. (2012). Antioxidant α-tocopherol checks lymphoma pro-
       motion via regulation of expression of protein kinase C-α and c-Myc genes and gly-
       colytic metabolism. *Leukemia & Lymphoma*, 1042-8194, Print, 1029-2403, Online, 53(6),
       1203-1210.

[121] Sharmila, R., & Manoharan, S. (2012). Anti-tumor activity of rosmarinic acid in 7,12-
       dimethylbenz(a) anthracene (DMBA) induced skin carcinogenesis in Swiss albino
       mice. *Indian Journal of Experimental Biology*, 50, 187-194, 0975-1009, Print, 0019-5189,
       Online.

[122] Zakaria, Z. A., Rofiee, MS, Mohamed, A. M., the, L. K., & Salleh, M. Z. (2011). In vitro
       antiproliferative and antioxidant activities and total phenolic contents of the extracts
       of Melastoma malabathricum leaves. *Journal of Acupuncture and Meridian Studies*, 4(4),
       248-256, 0000-0020.

[123] Sies, H. (1997). Antioxidants in Disease Mechanisms and Therapy. *Advances in Phar-
       macology*, 38, Academic Press: San Diego.

[124] Sies, H. (1997). Oxidative Stress: Oxidants And Antioxidants. *Experimental Physiology*,
       82, 291-295, 0958-0670, Print, 1469-445X, Online.

[125] Sivagami, G., Vinothkumar, R., Preethy, CP, Riyasdeen, A., Akbarsha, MA, Menon,
       V. P., & Nalini, N. (2012). Role of hesperetin (a natural flavonoid) and its analogue on
       apoptosis in HT-29 human colon adenocarcinoma cell line- A comparative study.
       Food and Chemical Toxicology, 50, 660-671, 0278-6915.

[126] Skalicky, J., Muzakova, V.,., Roman, Kandar. R., Meloun, M., Rousar, T., & Palicka, V.
       (2008). Evaluation of oxidative stress and inflammation in obese adults with metabol-
       ic syndrome. *Clinical Chemistry and Laboratory Medicine*, 46(4), 499-505, 1434-6621,
       Print, 1437-4331, Online.

[127] Slaga, T. J. (1995). Inhibition of the induction of cancer by antioxidants. *Advances in
       Experimental Medicine and Biology*, 369, 167-174, 0065-2598.

[128] Solanas, M, Hurtado, A, Costa, I, Moral, R, Menendez, J.A., Colomer, R., & Escrich, E.
       (2002). Effects of a high olive oil diet on the clinical behavior and histopathological
       features of rat DMBA-induced mammary tumors compared with a high corn oil diet.
       International Journal of Oncology, 1791-2423, 21, 745-753.

[129] Sun, Y, Yin, T, Chen, XH, Zhang, G, Curtis, RB, Lu, ZH, & Jiang, JH. (2011). In vitro
       antitumor activity and structure characterization of ethanol extracts from wild and

cultivated Chaga medicinal mushroom, Inonotus obliquus (Pers.:Fr.) Pilát (Aphyllophoromycetideae). *International Journal of Medical Mushrooms*, 1521-9437, Print, 1940-4344, Online, 13, 121-130.

[130] Surh, YJ. (2003). Cancer chemoprevention with dietary phytochemicals. *Nature Review Cancer*, 3, 768-780, 1097-0142, Online.

[131] Szpetnar, M., Matras, P., Kiełczykowsk,a, M., Horecka, A., Bartoszewska, L., Pasternak, K., & Rudzki, S. (2012). Antioxidants in patients receiving total parenteral nutrition after gastrointestinal cancer surgery. *Cell Biochemistry and Funciont*, 30, 211-216, 1099-0844, Online.

[132] Takada, M, Ku, Y., Habara, K, Ajiki, T., Suzuki, Y., & Kuroda, Y. (2002). Inhibitory effect of epigallocatechin-3-gallate on growth and invasion in human biliary tract carcinoma cells. *World Journal of Surgery*, 0364-2313, Print, 1432-2323, Online, 26, 683-686.

[133] Takada, M., Nakamura, Y., Koizumi, T., Toyama, H., Kamigaki, T., Suzuki, Y., Takeyama, Y., & Kuroda, Y. (2002). Suppression of human pancreatic carcinoma cell growth and invasion by epigallocatechin-3-gallate. *Pancreas*, 25, 45-48, 0885-3177, Print, 1536-4828, Online.

[134] Thapa, D., & Ghosh, R. (2012). Antioxidants for prostate cancer chemoprevention: Challenges and opportunities. *Biochemical Pharmacology*, 83, 1319-1330, 0006-2952.

[135] Toyoda-Hokaiwado, N, Yasui, Y, Muramatsu, M, Masumura, K, Takamune, M, Yamada, M, Ohta, T, Tanaka, T, & Nohmi, T. (2011). Chemopreventive effects of silymarin against 1,2-dimethylhydrazine plus dextran sodium sulfate-induced inflammation-associated carcinogenicity and genotoxicity in the colon of gpt delta rats. *Carcinogenesis*, 0143-3334, Print, 1460-2180, Online, 32, 1512-1517.

[136] Toyokuni, MD. (1998). Oxidative Stress and Cancer: The Role of Redox Regulation Shinya. *Biotherapy*, 11, 147-154, 0092-1299 X, Print, 1573-8280, Online.

[137] Tsaluchidu, S., Cocchi, M., Tonello, L., & Puri, B. K. (2008). Fatty acids and oxidative stress in psychiatric disorders. *BMC Psychiatry*, 8(1), S1-S5, 0147-1244 X, Online.

[138] Tumbas, V. T., Canadanović-Brunet, J. M., Cetojević-Simin, D. D., Cetković, G. S., Ethilas, S. M., & Gille, L. (2012). Effect of rosehip (Rosa canina L.) phytochemicals on stable free radicals and human cancer cells. *Journal of the Science of Food and Agriculture*, 92, 1273-1281, 0022-5142, Print, 1097-0010, Online.

[139] Upham, B. L., & Wagner, J. G. (2001). Toxicological Highlight Toxicant-Induced Oxidative Stress in Cancer. *Toxicological sciences*, 64, 1-3, 1096-6080, Print, 1096-0929, Online.

[140] Ursini, F., Maiorino, M., Morazzoni, P., Roveri, A., & Pifferi, G. (1994). A novel antioxidant flavonoid (IdB 1031) affecting molecular mechanisms of cellular activation. *Free Radical Biology & Medicine*, 16, 547-553, 0891-5849.

[141] Uttara, B., Singh, A. V., Zamboni, P., & Mahajan, R. T. (2009). Oxidative Stress and Neurodegenerative Diseases: A Review of Upstream and Downstream Antioxidant Therapeutic Options. *Current Neuropharmacology*, 7, 65-74, 0157-0159 X.

[142] Van Erk, M. J., Roepman, P., van der Lende, T. R., Stierum, R. H., Aarts, J. M., van Bladeren, P. J., & van Ommen, B. (2005). Integrated assessment by multiple gene expression analysis of quercetin bioactivity on anticancer-related mechanisms in colon cancer cells in vitro. *European Journal of Nutrition*, 44, 143-156, 1436-6207, Print, 1436-6215, Online.

[143] Valadez-Vega, C., Guzmán-Partida, A. M., Soto-Cordova, F. J., Alvarez-Manilla, G., Morales-González, J. A., Madrigal-Santillán, E., Villagómez-Ibarra, J. R., Zúñiga-Pérez, C., Gutiérrez-Salinas, J., & Becerril-Flores, MA. (2011). Purification, biochemical characterization, and bioactive properties of a lectin purified from the seeds of white tepary bean (phaseolus acutifolius variety latifolius). *Molecules*, 21, 2561-2582, 1420-3049.

[144] Valadez-Vega, C., Alvarez-Manilla, G, Riverón-Negrete, L, García-Carrancá, A, Morales-González, JA, Zuñiga-Pérez, C, Madrigal-Santillán, E, Esquivel-Soto, J, Esquivel-Chirino, C, Villagómez-Ibarra, R, Bautista, M, & Morales-González, A. (2011). Detection of cytotoxic activity of lectin on human colon adenocarcinoma (Sw480) and epithelial cervical carcinoma (C33-A). *Molecules*, 1420-3049, 2, 2107-2118.

[145] Varma, SD, Devamanoharan, S., & Morris, SM. (1995). Prevention of cataracts by nutritional and metabolic antioxidants. *Crititical Reviews in Food Science and Nutrition*, 35, 111-129, 1040-8398, Print, 1549-7852, Online.

[146] Vauzour, D, Rodriguez-Mateos, A, Corona, G, Oruna-Concha, MJ, & Spence, JPE. (2010). Polyphen ols and Human Health: Prevention of Disease and Mechanisms of Action. *Nutrients*, 2072-6643, 2, 1106-1131.

[147] Warburg, O. (1956). On the origin of cancer cells. *Science*, 123, 309-314, 0036-8075, Print, 1095-9203, Online.

[148] Waris, G., & Siddiqui, A. (2005). Hepatitis C virus stimulates the expression of cyclooxygenase-2 via oxidative stress: role of prostaglandin E2 in RNA replication. *Journal of Virology*, 79, 9725-34, 0002-2538 X, Print, 1098-5514, Online.

[149] Wayner, D. D. M., Burton, G. W., Ingold, K. U., Barclay, L. R. C., & Locke, S. J. (1987). The relative contributions of vitamin E, urate, ascorbate and proteins to the total peroxyl radical-trapping antioxidant activity of human blood plasma. *Biochemica et Biophysica Acta*, 924, 408-419, 0006-3002.

[150] Wei, L. S., Wee, W., Siong, J. Y., & Syamsumir, D. F. (2011). Charactetization of anticancer, antimicrobial, antioxidant properties and chemical composition of Peperomia pellucid. *Acta Medica Iranica*, 49, 670-674, 0044-6025, Print, 1735-9694, Online.

[151] Weijl, N. I., Cleton, F. J., & Osanto, S. (1997). Free radicals and antioxidants in chemotherapy induced toxicity. *Cancer Treatment Reviews*, 23, 209-240, 0305-7372, Print.

[152] Winslow, LC, & Krol, DJ. (1998). Herbs as medicines. *Archives Internal Medicine,* 0003-9926, Print, 1538-3679, Online, 1258, 2192-219.

[153] Wu, B, Li, J, Huang, D, Wang, W, Chen, Y, Liao, Y, Tang, X, Xie, H, & Tang, F. (2011). Baicalein mediates inhibition of migration and invasiveness of skin carcinoma through Ezrin in A431 cells. 1471-2407, 11, 527-536.

[154] Ye, F, Zhang, GH, Guan, BX, & Xu, XC. (2012). Suppression of esophageal cancer cell growth using curcumin, (-)-epigallocatechin-3-gallate and lovastatin. *World Journal of Gastroenterology,* 1007-9327, Print, 2219-2840, Online, 18, 126-135.

Chapter 4

Antioxidant Role of Ascorbic Acid and His Protective Effects on Chronic Diseases

José Luis Silencio Barrita and
María del Socorro Santiago Sánchez

Additional information is available at the end of the chapter

1. Introduction

Ascorbic acid (AA), commonly known as vitamin C, plays an important role in the human body, although its function at the cellular level is not yet clear. It is necessary for the synthesis of collagen, a protein that has many connective functions in the body. Among the substances and structures that contain collagen are bone, cartilage and the surrounding material, as well as carrier substances and materials of union muscle, skin and other tissues. It also requires (AA) for the synthesis of hormones, neurotransmitters and in the metabolism of certain amino acids and vitamins. Participate in the liver for detoxification of toxic substances and blood level for immunity. As an antioxidant reacts with histamine and peroxide for reducing inflammatory symptoms.

Its antioxidant capacity is associated with reduced incidence of cancer. The requirement for vitamin C for adults is well defined but they have not been uniform across different cultures, so their need has been defined as culture-specific. They have also defined other roles in cellular processes and reactions. Some epidemiological data mentioned its usefulness in reducing cold with increasing consumption of foods rich in vitamin, so people sometimes ingest an overdose of it. In most reports mention that discrete increases in blood levels of this vitamin reduces the risk of death in all conditions. Although there are many functions of vitamin C, his role in health is discussed mostly in relation to its role as an antioxidant and its effects on cancer, blood pressure, immunity, drug metabolism and urinary excretion of hydroxyproline.

Antioxidants play important roles in cellular function and have been implicated in processes associated with aging, including vascular, inflammatory damage and cancer. In the case of

AA its antioxidant role is useful since it contributes to the maintenance of the vascular system and the reduction of atherogenesis through regulation in collagen synthesis, production of prostacyclin and nitric oxide. In addition to this antioxidant role, the AA has actions at the molecular level because it acts as a cofactor of enzymes such as dopamine hydroxylase (EC 1.14.17.1), influencing neurotransmitter concentration, improves lysosomal protein degradation and mediates consumer monosodium glutamate.

1.1. History

Since the nineteenth and early twentieth century research on these compounds led to the discovery of vitamins.[4] Since 1901, a publication of Wildiers first described the stimulating effect of small amounts of organic material in the growth of yeast; this effect was the subject of many publications and only after several years was universally accepted. Wilders gave the name "bios" to the substance or substances causing increased growth of yeast. In the years since it was shown that bios were multiple in nature, and changes fractionated as bios I, bios IIA, IIB and others.[5]

1747. - Lind cured scurvy in British sailors with oranges and lemons.

1907. - It is reproduced experimental scurvy in guinea pigs.

1928. - Eascott identified bios I as meso-inositol. Szent-Gyorgy and Glen published the isolation of vitamin C or hexuronic acid.

1933. - Allison, Hoover and Burk describe a compound that promotes the respiration and growth of Rhizobium, which designated "Coenzyme R". Then, it is defined molecular structure and synthesis of vitamin C.

1937. - the Nobel Prize in Chemistry was awarded to Walter Haworth for his work in determining the structure of ascorbic acid (shared with Paul Karrer for his work on vitamins) and the Nobel Prize for medicine was awarded to Albert von Szent-Györgyi Nagyrápolt for his studies on the biological functions of ascorbic acid.

2. Definition

Vitamin C is defined as hexuronic acid, cevitáminic acid or xiloascórbic acid. The term vitamin C is generally used to describe all these compounds, although the representative of which is ascorbic acid.

2.1. Structure, formula and chemical characteristics

Ascorbic acid is the enolic form of one α-ketolactone. Ascorbic acid solution is easily oxidized to the diketo form referred to as dehydroascorbic acid, which can easily be converted into oxalic acid , diketogulonic acid or threonic acid.

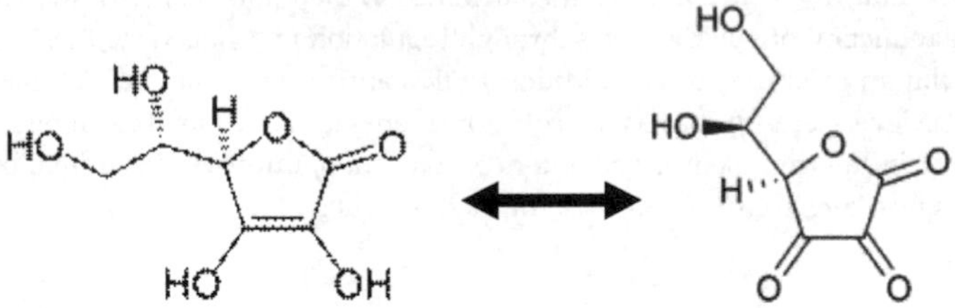

Figure 1. Structure of ascorbic acid and dehydroascorbic acid.

Name (IUPAC) systematic: (R) -3, 4-dihydroxy-5-((S) -1, 2-dihydroxyethyl) furan-2 (5H)-one; CAS Number 50-81-7: Formula: $C_6H_8O_6$; mol wt. 176.13 g / mol

2.2. Physical and chemical properties

Ascorbic acid contains several structural elements that contribute to their chemical behavior: the structure of the lactones and two enolic hydroxyl groups and a primary and secondary alcohol group. Enediol structure motivates their antioxidant properties, as can be oxidized easily enediols to diketones. Therefore, the carbonyl groups endioles neighbors are also called reductive.

Ascorbic acid forms two bonds intermolecular hydrogen bonds (shown in red in the figure) that contribute substantially to the stability and with it the chemical qualities of the structure endiol.

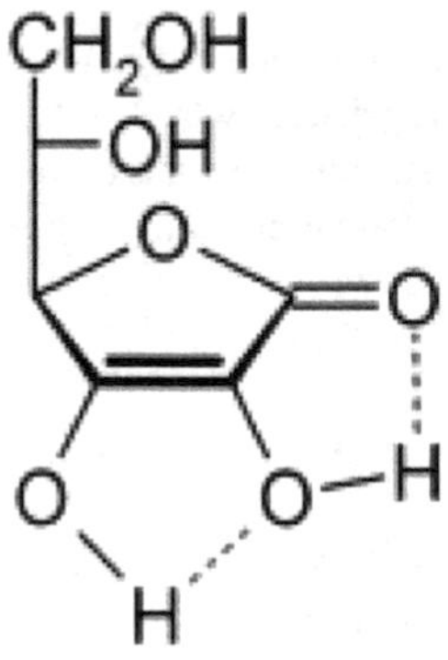

Figure 2. Hydrogen bridges formed by ascorbic acid

Ascorbic acid is rapidly interconvert in two unstable diketone tautomers by proton transfer, though it is most stable in the enol form. The proton of the enol is lost, and again acquired by the electrons from the double bond to produce a diketone. This reaction is an enol. There are two possible ways: 1, 2-diketone and 1, 3-diketone (figure 3).

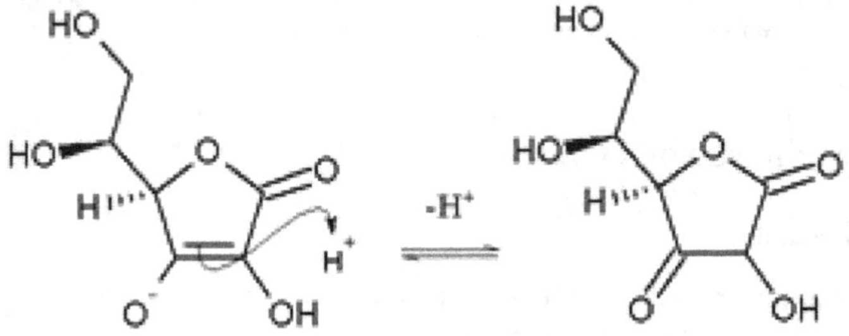

Figure 3. Nucleophilic attack on the proton ascorbic enol to give 1,3 diketone.

2.3. Vitamers or vitameric forms

The vitamer of a particular vitamin is any chemical compound which generally has the same molecular structure and each shows a different vitamin activity in a biological system which is deficient of the vitamin.

The vitamin activity of multiple vitamers is due to the ability (sometimes limited) of the body to convert one or many vitamers in another vitamer for the same enzymatic cofactor which is active in the body as the most important form of the vitamin. As part of the definition of the vitamin, the body can not completely synthesize an optimal amount of vitamin activity of foodstuffs simple, without a certain minimum amount of vitamer as base. Not all vitamers have the same vitamin power by mass or weight. This is due to differences in the absorption and the variable interconversion several vitamers in the vitamin.

For ascorbic acid per se, may be mentioned the following vitamers: dehydroascorbic acid, erythorbic acid (figure 4) and the following salts: sodium ascorbate, calcium ascorbate, and others. (Rogur, 2010)

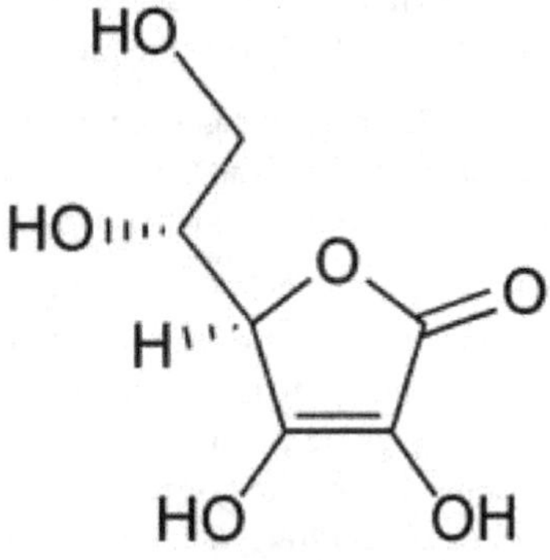

Figure 4. Structure of erythorbic acid, isoascorbic or Arabian-ascorbic

Chemical data: IUPAC Name: D-isoascorbic acid, CAS Number: [89-65-6] Molecular formula: $C_6H_8O_6$, molar mass: 176.13 g/mol; Melting point: 169-172 ° C

3. Relationship with other nutrients

3.1. Vitamin. A and E

A short-term supplementation with physiological doses of antioxidant vitamins, carotenoids and trace elements during alcohol rehabilitation clearly improves micronutrient status indicators. Heavy smokers in particular seem to respond to vitamin C supplementation (Gueguen, 2003)

In the liver accumulate micronutrients such as Vitamin A, E and iron, and therefore, in patients with hepatic impairment is a deficit of them due to reduced intake, as well as intestinal transport and liver stores. The alteration of fat soluble vitamins is especially important in patients with steatorrhea or cholestasis.

Moreover, in the alcoholic patient levels water-soluble vitamins are low due to the effect of ethanol on its metabolism, producing pyridoxine deficiency, retinol, cobalamin, folate and niacin. In fact, in chronic alcoholism may develop Wernicke encephalopathy. It has also shown a direct relationship between oxidative stress and disease severity liver, requiring the micronutrients with antioxidant activity, being increased the needs of vitamin E and C.

There are few data on vitamin needs and trace elements in burned patients. the large tissue loss, decreased gastrointestinal absorption, increased urinary losses, changes in distribution, and a high degree catabolism, that are made increased the needs of vitamins and trace elements. Clinical guidelines recommend giving also established daily requirements and additional doses of certain micronutrients.

Also, increase the intake of vitamin C (1,000 mg/day) as it promotes the healing process, and vitamin A (10,000 IU/day) for its effect immune and protective skin and mucous membranes. Also, is required additional vitamin D due to high risk of fractures in these group patients but have not yet been established Daily exact requirements.

Zinc supplement is suitable dose of 220 mg/day, as is involved in protein synthesis and tissue regeneration. Furthermore, Chan et al indicate that in the week post-injury, there are high losses exuding of copper, being necessary to increase their requirements (4.5 mg/day of copper sulfate).

The increased production of oxygen species reactive in this clinical situation requires administration of antioxidants (ascorbic acid, glutathione, carotenoids, vitamin A and E) been shown to reduce mortality, protecting the micro vascular circulation, reducing the peroxidation tissue lipid. According to some authors, surgical stress may necessitate supplementation of ascorbic acid, alpha tocopherol and trace elements, associating too low preoperative levels of vitamin A (<0.77 mol/L) with an increase of postoperative infection and mortality. At present, it is unknown whether supplementation micronutrient for a short period of time could restore plasma antioxidant levels after surgery. Some authors suggest that antioxidants could lead to improved metabolism and ventricular function after cardiac surgery. Also state that patient's major surgery may benefit from selenium, even before surgery, to action at the level of oxidative stress. The ESPEN recommended in these surgical patients treated with parenteral nutrition, supplement micronutrient recommended daily doses, the vitamin supplementation is unnecessary if the patient is on concomitant enteral nutrition, oral or parenteral.

3.2. Minerals such as selenium, iron and zinc

Recent studies have shown that in patients with inflammatory bowel disease there is a correlation between the level of some antioxidants such as selenium, vitamin C and E and clinical improvement and reduction in serum levels of TNF-α and decrease in steroid dose to 65%. Low serum zinc levels have been correlated with increased blood pressure, disease coronary type II diabetes mellitus and hyperlipidemia. Also, high intake of magnesium (> 500-1000 mg/day) can lower high blood pressure, and be effective in acute myocardial infarction and atherosclerosis. Houston recommended to prevent the emergence and development of hypertension, administration of additional vitamins and trace elements. Finally, oxidative stress plays an important role in the initiation and maintenance of the pathogenesis of cardiovascular disease and its complications. For some authors, antioxidants such as vitamins E and C, beta-carotene, selenium and zinc, may act by reducing the cardiovascular risk, although evidence is limited.

Malnutrition is a common feature of inflammatory intestinal diseases, being frequent the deficit of vitamins B_{12}, A, D, E and K, steatorrhea, ileal resection or extensive lesions in the intestine. GI bleeding contributing to iron losses, and through diarrhea and fistulas is loss of electrolytes and trace elements (copper, magnesium, selenium and zinc).

The role of antioxidant micronutrients in the clinical and functional improvement has been described by different authors. Thus, a low intake of selenium, beta-carotene and vitamins E and C, can reduce the body's natural defenses and increase inflammation of the airways, and therefore the selenium (100-200 ug/d) been associated with improved lung function, especially in smokers. Gazdik et al indicate that supplementation of 200 ug/d of selenium in asthmatic patients produced a statistically significant decrease in the use of corticosteroids. Loannidis and McClave et al indicate that antioxidants such as selenium, vitamin A, vitamin C and vitamin E reduce pancreatic inflammation and pain, and prevent the occurrence of exacerbations. Recently in a double blind study in patients with chronic pancreatitis were given daily supplements of 600 ug of selenium, 9,000 IU of beta-carotene, 540 mg of vitamin C, 270 IU of vitamin E and 2,000 mg methionine, the pain was reduced significantly, as well as stress markers oxidative normalized plasma concentrations of antioxidants.

For some authors, parenteral administration of ascorbic acid can lower the morbidity and mortality of these patients in a randomized, double-blind placebo-controlled; we observed that mortality at day 28 decreased in the group of patients who received ascorbate and vitamin E by intravenous infusion. Some authors recommend increasing the contribution of antioxidants such as vitamin C, retinol, vitamin E, beta-carotene and selenium. Also appear to require thiamine, niacin, vitamin A, E and C, B complex, zinc (15-20 mg/day and 10 mg/L intestinal leaks) and selenium (rise up to 120 ug/day) in patients with sepsis.

3.3. Polyunsaturated fatty acids

Consumption of 0.6 mg equivalents of alpha tocopherol/g linoleic acid is suitable for human adults. The minimum requirement of vitamin E related to the consumption of fatty acids with a high degree of unsaturation can be calculated with a specific formula that must take into account the peroxibility of polyunsaturated fatty acids is based on the results of animal experiments. But still no convincing evidence of how much vitamin E is required in relation to consumption of polyunsaturated fatty acids: EPA (20:5 n-3) and DHA (22:6 n-3).

Studies so far show that the effects of supplementation with EPA and DHA lipid peroxidiza-cion increase even when the amounts of vitamin E present are adequate in relation to the oxida-tive potential of these fatty acids. On the other hand the calculation of the requirement for vitamin E using current data from recent consumption, show that a reduction in total fat intake with a concomitant increase in consumption of polyunsaturated fatty acids, including EPA and DHA results in an increased need intake of vitamin E. In fact the methods used to investigate the requirements of vitamin E and polyunsaturated fatty acid intake (erythrocyte hemolysis) and the techniques used to assess lipid peroxidation (malondialdehyde analysis, MDA) may be inappropriate for measuring a quantitative relationship between the two loads.

Therefore, further studies are needed to establish the requirement of vitamin E when intake of unsaturated fatty acids of longer chain increases. For this purpose it is necessary to use functional techniques based on the measurement of in vivo lipid peroxidation. Until then or until the available data suggest using the index of 0.6 mg of alpha tocopherol per gram of ingested PUFA. However it is likely that higher levels are necessary for vitamin fats are rich in fatty acids containing more than two double bonds (Valk, 2000).

The diet of our ancestors was less dense in calories, being higher in fiber, rich in fruits, vege-tables, lean meat and fish. As a result, the diet was significantly lower in total fat and satu-rated fat, but containing equal amounts of n-6 essential fatty acids and n-3. Linoleic acid (LA) is the major n-6 fatty acid and alpha-linolenic acid (ALA) is the major n-3 fatty acid.

In the body, LA is metabolized to arachidonic acid (ARA), and ALA is metabolized to eico-sapentaenoic acid (EPA) and docosahexaenoic acid (DHA). The essential fatty acid ratio n-6: n-3 is 1 to 2:1, with higher levels of long chain polyunsaturated fatty acids (PUFA), such as EPA, DHA and ARA, today's diet. Today this ratio is about 10 to 1 or 20 and 25 to 1, indicat-ing that Western diets are deficient in n-3 fatty acids compared with the diet that humans evolved and established patterns genetic.

The n-3 and n-6 are not interconvertible in the human body and are important components of practically all cell membranes. The n-6 fatty acids and n-3 influence eicosanoid metabo-lism, gene expression, and intercellular communication cell to cell. The polyunsaturated fat-ty acid composition of cell membranes is largely dependent on food ingestion. Therefore, appropriate amounts of n-6 fatty acids and n-3 in the diet should be considered in making dietary recommendations.

These two classes of polyunsaturated fatty acids must be distinguished because they are metabolically and functionally distinct and have opposing physiological functions, balance is important for maintaining homeostasis and normal development. Studies with nonhuman primates and human newborns indicate that DHA is essential for normal functional development of the retina and the brain, especially in premature babies. A balanced n-6/n-3 ratio in the diet is essential for normal growth and development and should lead to reduced cardiovascular disease and other chronic diseases and improve mental health. Although a recommended dietary allowance for essential fatty acids does not exist, an adequate intake (AI) has been estimated for essential fatty acids n-6 and n-3 by an international scientific working group. The final recommendations are for Western societies, reduce the consumption of n-6 fatty acids and increased intake of n-3 fatty acids. (Simopoulus, 2000)

4. Vitamin C sources

4.1. Food sources in Mexico

The main sources of ascorbic acid are presented in Table 1. The results of vitamin C are shown as the mean and correspond to the official tables of composition of Mexican foods.

FOOD	CONCENTRATION (mg/100g EP)
poblan chili	364
"trompito" chili	320
yucca flower	273
guava	199
Marañon	167
cauliflower	127
red bell pepper	160
guajillo chili	100
garlic	99
Nanche fruit	86
orange	76
manila mango	76
serran chili	65
pumpkin	58
watercress	51
beet	20
cucumber	13

EP = edible portion

Table 1. Main sources of ascorbic acid in México

4.2. Fruits and vegetables

Vitamin C is a major constituent of fruits and vegetables, which also contain citric acid, oxalates and substances such as anthocyanins, coloring agents and carotenoids that are difficult to quantify when using colorimetric methods.

Currently there is great interest in relation to consumption of natural foods and mainly on the content of nutrients in fruits, vegetables and vitamin C. This interest is due in part to vitamin C is probably one of the most widely used nutrients in the food and pharmaceutical industry. Used as a supplement, additive, preservative, as an antioxidant in processed foods. Table 2 shows the main foods that are good sources of vitamin C.

FOOD	CONCETRATION (mg/100g)
Acerola	1743
Coriander leaves dry	567
red peppers, spicy, ripe, raw	369
Orange juice, dehydrated	359
Grape juice, dehydrated	350
common Guavas, whole, raw	242
Dried Tomato juice	239
red pepper, spicy, immature, raw	235
Lemon juice	230
Orange juice, canned	229
Col common, dehydrated	211
Peppers, sweet, ripe, red, raw	204
Currants, black, European, raw	200
parsley, dry	172
Orange juice	144
green radish, raw	139
Grape juice	138
Orange peel, raw	136
Pokeroot fruit	136
Mustard seeds	130
leaves of kale, boiled, drained	93
Broccoli cooked	90
Brussels sprouts cooked	87
Lamb crude	80
Leaves and stems of cress, raw	79
Cauliflower, raw	78
Brussels sprouts, cooked in water	76
Cabbage, red, raw	61
Strawberries Raw	59
Papaya	56

Table 2. Acid content of ascorbic acid in different foods and different presentations

In Table 3 shows the analysis of vitamin C in different parts of the food such as edible portion thereof, the seed or plant center and the shell and stalks that are normally discarded.

Food	Edible portion	Shell	Seed	Stem
Apple	458.1±17.8 (84.5)	848.9±11.2 (81.8)	--	--
chayote	7.2±0.9 (94.2)	19.9±0.5 (94.2)	266.6±6.7 (90.5)	--
onion	17.0±0.7 (87.6)	--	--	456.6±1.8 (93.0)
lime	256.2±17.2 (87.1)	1916.5±186.7 (69.5)	--	--
pomarrosa	531.7±36.4 (83.6)	623.0±35.1 (82.5)	1044.5±50.9 (65.7)	--
guanábana	140.7±3.6 (80.0)	82.1±4.9 (81.8)		
beat	70.8±6.8 (85.3)	115.8±1.2 (85.8)		
Potato	74.6±6.7 (80.3)	4.2±1.0 (85.8)		

Table 3. Distribution of ascorbic acid (mg/100g) in some fruits and vegetables produced in México

Below the edible portion the moisture (%) is indicated in parenthesis. The data shown in Tables 3 through 6 are original and have not been published yet. EP = Edible portion

Table 4 shows the concentration of ascorbic acid in the main fruits and vegetables consumed in Mexico. Data are presented as mean and standard deviation. It shows that even the concentration of vitamin C is lower in the edible portion in the shells reported in Table 5.

Food	Concentration (mg / 100 g EP)	Moisture (%)
squash	7.2±0.9	94.2
spinach	8.5±0.2	91.6
potatoes	74.6±6.7	79.7
cucumber	93.0±7.1	96.0
green tomato	222.8±10.7	90.9
poblano chile	191.0±7.8	91.9
green pepper	195.5±9.5	94.3
nopales	268.9±30.1	95.2
cambray onion	17.0±0.7	93.0
carrot	50.4±5.6	87.2
white cabbage	184.7±17.2	93.2
grapefruit	261.3±10.7	87.8
mango	319.6±5.3	84.6
watermelon	56.2±8.9	92.5
banana	333.7±6.3	76.5
orange	279.8±39.4	84.0
mamey	31.6±1.2	66.8
plum	331.1±17.9	88.1
grape	66.1±13.6	84.3
apple	458.1±17.8	84.5
beet	70.8±6.8	85.3
lemon	39.4±2.5	88.3
avocado	256.2±61.9	84.3
sweet lime	306.8±23.4	89.8

Table 4. Content of ascorbic acid in Mexican fruits and vegetables

The shell or skin of many fruits and vegetables is usually discarded, but these same wastes have significant amounts of ascorbic acid, which is shown in Table 5, where the lime peel, rose apple and apple are the best examples.

Shell of	concentration (mg/100g)	moisture (%)
CARROT	129.1±10.1	88.5
APPLE	848.9±11.2	81.8
BITTER LEMON	39.4±2.5	71.7
AVOCADO	187.1±27.2	82.0
BEET	115.8±1.2	85.8
SMALL POTATO	4.2±1.0	85.5
BIG POTATO	2.1±0.1	80.3
POMARROSA	622.9±35.1	82.5
LIME	1916.4±186.7	69.4
CHAYOTE	19.9±0.5	94.2

Table 5. Content of ascorbic acid in the shell of some fruits and vegetables

Table 6 shows the values of ascorbic acid in some plant species used as flavoring for Mexican dishes. In most cases the amount used for the preparation of food is very low and sometimes do not amount to more than 2% by weight of the end plate. However their presence in cooked food gives organoleptic properties suitable for the acceptance of it and especially the potential of the flavors of food.

spices	concentration (mg /100g EP)	moisture (%)
coriander	56.0±7.3	85.3
yerbabuena	226.9±21.9	86.8
epazote leaves	18.3±1.1	86.3
chard leaves	8.5±0.9	92.7
chard stem	15.7±0.9	95.0
parsley leaves	243.9±17.2	84.5
parsley stem	182.1±11.0	90.2
celery leaves	5.6±0.2	88.4
celery stem	7.8±0.9	90.3
corn grain	35.3±6.0	70.7
corn "hair"	55.9±1.2	85.4

Table 6. Content of ascorbic acid in some food used as spices

Food-industrialized

Vitamin C (ascorbic acid) is water soluble and sensitive to oxygen. For this reason, it may partially destroyed in foods during processing, if exposed to air during storage or if treated with water. Manufacturers can protect them from oxidation by adding vitamin C. The addition of ascorbic acid as an antioxidant should be appropriately marked in the list of ingredients on the label of the final product

Ascorbic acid and its salts are practically insoluble in lipids (fats), for this reason that is often used in the food industry as an antioxidant and preservative greasy foods, in order to avoid rancid. Their salts are usually used with a solubilizing agent (usually a monoglyceride) to improve its implementation. Also this often used in the processing industry of cosmetics products. Sodium ascorbate is a sodium salt of ascorbic acid (vitamin C) and formula $C_6H_7NaO_6$. This form is used in the food industry for their functions antiseptic, antioxidants, and preservatives. Ascorbyl palmitate is an ester formed by ascorbic acid (vitamin C) and palmitic acid creating a liposoluble form of vitamin C. It is used in the food industry as an antioxidant (code E 304). It is wrong to think that is a natural antioxidant

Use as a preservative

Is usually used as a food preservative and as antioxidant in the food industry, a typical case is found as a bread improver additive. In industry collecting fruit prevents the color oxidative change known browning. Is often added to foods treated with nitrite in order to reduce the generation of nitrosamines (a carcinogen), so commonly found in sausages and cold cuts. In the same way is generally used in the food industry as an acidity regulator.

Ascorbic acid and its sodium, potassium and calcium salts are used widely as antioxidants and additives. These compounds are soluble in water, so that fats do not protect against oxidation. For this purpose may be used ascorbic acid esters with fat soluble long chain fatty acids (palmitate and ascorbyl stearate).

5. Deficiencies

5.1. Primary deficiency – scurvy. Signs and symptoms

A frank deficiency of vitamin C causes scurvy, a disease characterized by multiple hemorrhages. Scurvy in adults is manifested by lassitude, weakness, irritability, vague muscle pain, joint pain and weight loss. Early signs objectives are as bleeding gums, tooth loss and gingivitis.

The diagnosis of scurvy, is achieved by testing plasma ascorbic acid, low concentration indicates low levels in tissues. It is generally accepted that ascorbic acid concentration in the layer of coagulated lymph (20-53 ug/10^8 leukocytes) is the most reliable indicator of nutritional status regarding vitamin C and its concentration in tissues. In the most extreme cases scurvy shows: bleeding gums and skin, perifollicular bleeding and ocular petechiae, salivary and lachrymal glands "drier", functional neuropathy, lower limb edema, psychological disturbances, anemia and poor healing of wounds. The consumption of snuff lowers blood levels of vitamin C.

Clinical manifestations of Vitamin C deficiency

Clinical manifestations of deficiency have been described at several levels: a) mesenchymal. - By the presence of petechiae, ecchymosis, curly hair, peri-follicular hemorrhages, bleeding gums, swollen, hyperkeratosis, Sjôgren syndrome, dyspnea, arthralgia, edema and poor healing, b) systemics, characterized by fatigue, weakness and lassitude c) psychological and neurological, by the presence of depression, hysteria, hypochondriasis and vasomotor instability.

5.2. Deficiencies secondary and association with other diseases

Severe deficiency of vitamin C leads to Scurvy. This rarely occurs but can be observed deficiencies in those who consume a diet without vegetables and fruits, alcoholism, in older people with limited diets, severely ill patients with chronic stress and in infants fed cow's milk.

Symptoms of scurvy are follicular hyperkeratosis, gingival swelling and inflammation (in gums), bleeding gums, loose teeth, dry mouth and eyes, hair loss and dry skin, among other symptoms that can lead to death.

By deficiency of collagen, the wounds do not heal scars and wounds of previous rupture and may lead to secondary infections. Neurotic disorders are common, consisting of hysteria and depression, followed by decreased psychomotor activity. It is not safe indiscriminate administration of ascorbic acid, since as the body becomes saturated, decreased absorption, and giving large doses, abruptly deleted. So if you continue with diet low in vitamin, may appear "rebound scurvy". In addition to "rebound scurvy," gastric intolerance and kidney, its use decreases the cobalamin (vitamin B_{12}), a substance synthesized by the body.

Eating a balanced and varied diet high in fruits and vegetables, the minimum dose of vitamin C, this completely covered. The daily requirement in an adult male is 90 mg/d and a woman of 75 mg/d (mg/day), although there are always situations where it is necessary to increase the dose of vitamin A through supplementation. Such circumstances or situations are:• Pregnancy and Lactation, • Alcoholics and smokers, • diabetics, • Allergic and asthmatic, • People who take daily medications or medications such as oral contraceptives, cortisone, antibiotics, etc.

Anemia by Vitamin C Deficiency

Anemia of vitamin C Deficiency is a rare type of anemia that is caused by a severe and very prolonged lack of vitamin C. In this type of anemia, the bone marrow produces small red blood cells (microcytosis). This deficiency is diagnosed by measuring the values of vitamin C in white blood cells. One tablet of vitamin C per day corrects the deficiency and anemia cure.

Associated symptoms

Associated deficiency or lack of vitamin C (ascorbic acid) can produce or be reflected by: Swollen and bleeding gums, dry, rough skin, spontaneous bruising, Impaired wound healing, bleeding nose, joint pain and swelling, anemia and dental enamel weakened. Very small amounts of vitamin C may be associated with signs and symptoms of deficiency, including: • Anemia, • Bleeding gums, • Decreased, ability to fight infection. • Decrease the rate of

wound healing. • Dry and separated strands in the hair. • Tendency to bruising. • Gingivitis (gum inflammation). • nosebleeds. • Possible weight gain because of slowed metabolism.
• Rough, dry and scaly. • Pain and swelling of the joints. • Weakened tooth enamel.

6. Laboratory methods for its measurement in foods and biological fluid

In the analysis of vitamin C, for the methods commonly used consume many time and
therefore overestimates the concentration, due to other oxydizable species different of vitamin C; the determination by liquid chromatography of high resolution (HPLC), with electrochemical detection, for example, they require equipment not always available in smalls
laboratories and also, is very expensive. However this method quantifies all the forms of the
vitamin C present in the sample, and even it detects an epimer of ascorbic acid, the eritorbic
acid or isoascorbic acid. The samples of vitamin C saturation are used to establish the deficiency of ascorbate in tissue and are useful to confirm the diagnostic of scurvy when the patient has a normal absorption (Engelfried, 1944). It has been described 3 types of tests to
determine the tissue saturation, the first 2 are easy to make but they don't cover the problem
on totally, the third test is complicated and it's only useful in research work listed below:

a. Measurement of blood levels with and without a test sample:

    • The vitamin C in the plasma is not found doing a metabolic function; it is rather in a
      transit

    • from one tissue to another. Its lost or decrease does not indicates the intracellular sta-
      tus of this vitamin. A well-nourish adult with a free acid ascorbic diet decreases his
      serum levels of acid to cero in about 6 weeks; however only after many weeks of
      more deprivation the scurvy symptoms appears. So that for this reason the scurvy
      patients have low levels of ascorbic acid in the plasma. The vitamin C determination
      in plasma after a charge dose generally reflects the vitamin proportion, which has
      been stored by the tissues, however is tough to do a completely quantitative techni-
      que, because in high doses of this vitamin, the plasmatic concentration exceeds the
      kidney threshold which causes the lost of this vitamin in urine. For a specific meas-
      ure most be given multiple small doses, to avoid an excess in the blood levels above
      the renal threshold.

b. Measurement of kidney excretion with and without sample dose: The most important
    problems in this measurement are those concerning to the collection of urine, more than
    the vitamin measurement. The excretion of the vitamin has been correlated with the cre-
    atinine excretion. This is because the creatinine is used as a real and simple indicator of
    glomerular filtration.

c. Tissue Measurement This is the one of the 3 techniques which gives a real representa-
    tion of desaturation of the vitamin. This measurement is difficult because of problems
    in the sampling of tissue. In the case of ascorbic acid is recommended two methods to
    measure tissue levels. In the first method is measured its concentration in the buffy coat

and platelets which correlates good with the first signs of scurvy, making the most recommended technique. The second method determines the tissue saturation grade for an
intradermal test, using dichlorophenol indophenol, which depends of skin reductor
substances, which made this nonspecific.

6.1. Spectrophotometric method

Another proposed method is highly sensitive colorimetric determination of ascorbic acid
with 2,4-dinitrophenylhydrazine.

The method can be applied in virtually any laboratory is simple to perform and requires
little complicated equipment compared to HPLC. Samples can be from serum to food. In
foods the determination may be affected if the food contains natural dyes interfering
reading the wavelength of detection. However, it is an easy to implement, since it is in-
expensive and sensitive.

The use of UV-Visible spectroscopy (UV-Vis), for the determination of AA, is widely used in
research with food, since this acid has strong electronic transitions in the UV region, facilitating their identification and quantification by this technique.

Vitamin C is easily affected by such factors as moisture, light, air, heat, metal ions such as
iron and copper, oxygen and the alkaline medium. After decomposition under these conditions is easily transformed into various compounds such as: oxalic acid, L-threonic acid, L-
xylonic acid and L-dehydroascorbic acid, and in turn the latter are irreversibly transformed
into acid-diketo 2.3 L gluconic acid, which is its main degradation product.

Within the official methods described for the analysis of vitamin C in tablet, one of the most
widely used is the direct titration with 2,6-dichlorophenol indophenol by simple and rapid
result. The method is valid if it is known that the composition of the sample no interfering
substances and the concentration of dehydroascorbic acid is negligible, therefore, can be applied to a freshly prepared sample, but not useful in stability studies of vitamin C.

6.2. Chromatographic method (HPLC)

However, high performance liquid chromatography (HPLC) ensures detection and quantita-
tion limits lower, which also facilitate the elimination of the effects caused by the matrix (interference to other methods of analysis); this technique used as an essential tool in detailed
kinetic studies.

Quantification of Ascorbic Acid by HPLC.

Using a standard solution of 50 mg AA/L phosphoric acid, 0.05 N and taking into account
the various reports, on the conditions for the quantification of AA in samples, it optimizes
the mobile phase and the wavelength of the detector (scanning from 200 to 320 nm) to ob-

tain the greatest sensitivity and resolution in the chromatographic signal. These tests are performed with the following mobile phases:

2% KH_2PO_4, pH = 2.3

Acetonitrile-Water (70:30)

1% NaH_2PO_4, pH = 2.7

Methanol-buffer solution: 0.03 M KH_2PO_4, pH = 2.7, (99:1)

Water-methanol-acetonitrile (74.4: 25.0: 0.6)

The test is performed with the standard addition curve and calibration curve AA patterns between 1.0 to 25.0 mg/L phosphoric acid, 0.05 N, to determine the effect matrix in the method of quantification. The proposal is a column Hypersil ODS C_{18}, 5μm x 4.0 mm x 250 mm.

6.3. Other methods reported

Iodometric titration

Ascorbic acid or Vitamin C ($C_6H_8O_6$) can be determined by means of an iodometric titration. Vitamin C is a mild reducing agent that reacts rapidly with tri-iodide ion, this reaction is generated in a known excess of tri-iodide ion (I_3-) by reacting iodate iodide, is allowed to react and then the excess is titrated by I_3-back with a solution of thiosulphate. -The method is based on the following reactions:

$$8I^- + IO_3- + 6H^+ \ ------ \rightarrow 3H_2O + 3I_3$$

$$C_6H_8O_6- + H_2O \ + \ I_3------ \rightarrow C_6H_8O_6 \ + \ 2H^+ + 3I$$

Ascorbic acid   dehydroascorbic acid

$$I^{3-} + 2(S_2O_3)^{-2} \ ------ \rightarrow 3I^- + (S_4O_6)^{-2}$$

thiosulfate   tetrathionate

7. Requirements and recommendations in Mexico

The recommended daily ingestion (intake) is of 60 to 100 mg to avoid the appearance of disease symptoms that are produced by deficiencies of this vitamin. The infants require a little more of 100mg/day, although there is controversy over the minimum amount of this vitamin. We must take into account that this vitamin is very labile at heat and oxygen presence. The ascorbic acid is specific in the treatment of scurvy; the required dose could be better measured by the urinary excretion after a saturation dose. Depending of the required saturation velocity is the daily dose recommended which varies between 0.2 and 2.0 g/d. In the vitamin C deficiency, the tissue saturation is obtained with 3 daily doses of 700 mg each one for 3 days. Harris and cols. defined the tissue saturation as a sufficient storage of ascorbic acid where occurs a excretion of 50 mg or even more in a period of 4 to 5 hours after 1 dose of 700 mg/d.

The decreased levels in smokers are basically explained because they consume fewer sources of the vitamin. In this kind of population will be required a 50% more of the recommended dose of the vitamin. Because of the daily recommendation (RDA) it is defined as the daily ingestion average of food which is enough to cover the nutriments required for healthy people in a group of the population, it's necessary to continually assess these recommendations for vitamin C. The totalities of the reviewed information suggest that a consumption of 90-100 mg of this vitamin is enough for the optimum reduction of chronic disease risk in non-smoking men and women. Although some reports are suggested amounts up to 120 mg/day.

8. Toxicity and hypersensitivity

High doses of the vitamin (5-15 g/day), may cause osmotic diarrhea because it is ingested more vitamin of which can be absorbed. Also ascorbic acid can provoke intestinal cramps and acidification of the urine, leading to the formation of oxalate stones in the kidney of urinary tract. An exaggerated complementation during pregnancy may high the fetal requirement and result in the presence of scurvy in the newborn. It is also credited with the destruction of vitamin B_{12} of food during the ingestion.

Since the oxalic acid is a metabolite of the catabolism of the ascorbic acid is likely to be the formation of oxalate crystals in kidney in patient's susceptible to nucleation and therefore the formation of crystals or "kidney stones" when it is consumed excess of the vitamin. This relation however does not extend to subjects which are not susceptible to the formation of these kidney stones.

9. Biochemical functions

9.1. Paper as an antioxidant

Vitamin C is a soluble antioxidant important in biological fluids. An antioxidant is defined as "any substance which, when present in lower concentrations compared with the oxidable substrates (for example, proteins, lipids and carbohydrates and even nucleic acids) avoids or prevent significantly the oxidation of this substratum". The definition also given by the Food Nutrition Board is "A dietary antioxidant is a substance present in food which decreases significantly the adverse effects of the reactive species of oxygen (ROS), reactive species of nitrogen (RNS), or both for the normal physiology function in humans.

9.2. Interaction with ROS

Vitamin C quickly debug the reactive species of oxygen and nitrogen just as superoxide, hidroperoxile radicals, aqueous peroxyl radicals, singlet oxygen, ozone, peroxynitrite, nitrogen dioxide, nitroxide radicals and hypochlorous acid thereby protecting in fact other substrate of the oxidative damage[16].

Although the AA (ascorbic acid) reacts quickly with the hydroxyl radicals (constant speed > 10^9 $Lmol^{-1}s^{-1}$) is clumsy to debug this radical preferentially over other substrates. This is because hydroxyl radicals are very reactive and they will combine immediately with nearest substratum in their environment at a limited speed because of its diffusion. Vitamin C can also act as a co-antioxidant when regenerate the α-tocopherol (vitamin E) from the α-tocopheroxil radical produced when this is debugged from the lipid-soluble radicals just made. This is a function potentially important because in the *in vitro* experiments have shown that α-tocopherol can act as a pro-oxidant in absentia of co-oxidants just as vitamin C. However the relevance *in vivo* of the interaction of both vitamins it's not that clear yet.

AA can regenerate urates, glutathione and β-carotene *in vitro* from their respective oxidation products with one unpaired electron (urate radicals, glutathionil radicals and cations of β-carotene radicals). Two important properties of vitamin C make it an ideal antioxidant. The first one is its low potential reduction of ascorbate (282 mV) and its oxidation product with an electron, the ascorbile radical (2174 mV), which is derivates from its functional group en-diol in the molecule. This low potential of reduction of the ascorbate and the ascorbile radical makes them potentially appropriate for oxidation-reduction reactions and that why the vitamin acts as a soluble antioxidant terminal molecule. The second property which makes the vitamin an ideal antioxidant is the stability and the low reactivity of the just made ascorbile radical when the ascorbate debug the reactive species of oxygen and nitrogen. (equation 1).

The ascorbyl radical disproportionate rapidly to form ascorbate and dehydroascorbic acid (equations 1 and 2), or it is retro-reduced to ascorbate by an enzyme semi-dehydroascorbate reductase dependent of NADH. The oxidation product of two AA electrons, dehydroascorbic acid, can be reduced by itself to ascorbate because of the glutathione, by enzymes dependents of glutathione: glutaredoxin (dehydroascorbate oxidoreductase [glutathione dehydrogenase (ascorbate)]) or by an enzyme dependent of selenium (seleno-enzyme): the tioredoxine reductase. Alternatively, the dehydroascorbic acid gets quickly and irreversibly hydrolyzed to 2,3-dicetogulonic acid (ADCG) (equation 3):

$$AH_2 \leftrightarrow A\bullet_2 \leftrightarrow A \tag{1}$$

$$A\bullet_2 + A\bullet_2 \rightarrow AH_2 + A \tag{2}$$

$$A \rightarrow ADCG \rightarrow \text{oxalate, threonate, etc.} \tag{3}$$

Where in equation 1 shows the reversible oxidation of 2 ascorbate electrons (AH_2) to the ascorbile radical ($A\bullet_2$) and dehydroascorbic acid (A) respectively; the equation 2 shows the dismutation of the ascorbile radical to transform ascorbate and dehydroascorbic acid; and equation 3 show the hydrolysis of the dehydroascorbic acid to ADCG, which decomposes to oxalate, treonate y many other products. Vitamin C has been recognized and accepted by the FDA as one of 4 dietary antioxidants, the other 3 are vitamin E, β-carotene precursor of vitamin A and selenium as an essential component of antioxidant enzymes glutathione peroxidase (GPx) and thioredoxin reductase.

Although there is substantial scientific evidence of the role of antioxidant vitamin C and its effects on human health are needed more research that guarantee the role of vitamin both *in vivo* and *in vitro*, particularly because the AA is a redox active compound, which may act not only as an antioxidant but also as pro-oxidant in the presence of ion redox active transition metal.

The reduction of metallic ions like iron and copper for the vitamin C *in vitro* (equation 4) results in the formation of hydroxyl radicals highly reactive way to the reaction of this ions with hydrogen peroxide, a process known as the Fenton chemistry (equation 5), The lipid hydroperoxides can also "break" because of reduced metallic ions, forming lipid alkoxy radicals (equation 6) which can begin and spread chain reactions of the lipidic peroxidation. However the shown mechanism in the equation 5 requires the availability of free ions, redox active metallic ions and a low index vitamin C/metallic ion, conditions unlikely under normal conditions *in vivo*. Although has been shown that in biological fluids like plasma, the vitamin C acts like an antioxidant towards the lipids even in presence of free active ions.

$$AH_2 + M(n+1) \rightarrow A\bullet_2 + Mn + H+ \tag{4}$$

$$H_2O_2 + Mn \rightarrow \bullet OH + OH_2 + M(n+1) \tag{5}$$

$$LOOH + Mn \rightarrow LO\bullet + OH_2 + M(n+1) \tag{6}$$

Where in equation 4 shows the reduction of metallic active redox ions $M(n+1)$] because of the ascorbate to form de ascorbile radical and the reduced metal (Mn), the equation 5 shows the productions of hydroxyl radicals highly reactive $(\bullet OH)$ of the reaction of the hydrogen peroxide (H_2O_2) with the reduced metallic ions and the equation 6 shows the reaction of the lipid hydroperoxides (LOOH) with the reduced metallic ions to form alkoxy radicals $(LO\bullet)$. Although there no convincing evidence of a prooxidant effect of vitamin C on humans, exist a substantial evidence of its antioxidant activity. Interestingly its antioxidant activity does not correlate directly in its anti-curvy effect.

Because of this, the experts considerate that if the antioxidant activity of Vitamin C is accepted *in vivo* and that if this is relevant for human health, then scurvy should not be considerate as the only criterion for the nutritional fitness or for determine the ideal quantity or required of the vitamin.

9.3. Molecular mechanisms intracellular

9.3.1. As an enzyme cofactor

The molecular mechanisms of the anti-scurvy effect of vitamin C are very broad and so low studied. Also Vitamin C is a cofactor of many involved enzymes in the collagen biosynthesis, carnitine and neurotransmitters. The pro-collagen dioxygenase (proline hydroxylase) and the procolagene-lisine-5-dioxigenese are two enzymes involved in the synthesis of the collagen which needs ascorbic acids for maximum activity. The posttranslational hydroxylation of the lysine and proline residues of these enzymes are indispensable for the synthesis and formation of the stable helix which forms the collagen. So that the difference of the vitamin leads to the formation of weak structures causing lost of teeth, pain in joints, disorders of the connective tissue, and poor healing, characteristic signs of scurvy.

Two dioxigenases involved in the carnitine synthesis also require vitamin C for its activity. Carnitine is essential for the transport of long chain fatty acids to the mitochondria so one deficiency of vitamin C will bring consequences just as fatigue and lethargy which are late symptoms of scurvy. Besides that the vitamin C is a cofactor for the synthesis of catecholamines, in particular for the conversion of dopamine to norepinephrine catalyzed by the enzyme dopamine-β-monooxygenase. Depression, hypochondriasis, and behavioral changes are common in scurvy as a result of deficient dopamine hydroxylation.

Other kind of enzymes where vitamin C acts as a cofactor are the ones involved in the peptides amidations and in the tyrosine metabolism (this are also of the mono and dioxygenases kind). It is also implicated in the cholesterol metabolism to bile acids, way of 7-α-monooxygenase and in the adrenal steroids metabolism. The hydroxylation of aromatic drugs and carcinogens by cythocrome P-450, gets better also by reducing agents like vitamin C.

The role of vitamin C, due to its redox potential is to reduce metal ions present in the active sites of enzymes mono and dioxygenases. Ascorbate for instance acts as a co substrate in these reactions, not as a coenzyme. The reduction of iron, involved by the presence of vitamin improves the intestinal absorption of dietary non heme iron. Other proposals include the maintenance of the thiol groups of proteins, keeping in its reduced form of glutathione addition, a cellular antioxidant and enzyme cofactor, and tetrahydrofolate as a cofactor required for the synthesis of catecholamine.

10. Ascorbic acid in cancer

The recommended daily allowance (RDA) for ascorbic acid varies from 100 to 120 mg/day for adults. Have been attributed many benefits just like its antioxidant power, antiatherogenic, anticarcinogenic, immunomodulatory and anti-cold. However these benefits have been subject of debate and controversies because of the danger in the use of mega doses often used and its prooxidant effects and antioxidants. Discussed even if ascorbic acid cause cancer or promote or interfere with cancer therapy, the experts panels of dietary antioxidants and related compounds have been concluded that the data *in vivo* does not shows clearly a direct relation between the excess ingestion and the formation of kidney stones, the prooxidant effects and the excess absorption of iron.

The epidemiological and clinic study does not shows conclusive benefic effects in many kinds of cancer, with the exception of stomach cancer. Recently it has tested several derivatives of ascorbic acid on cancer cells as ascorbic acid spheres. The ascorbyl stearate is a compound which inhibits the human carcinogenic cell proliferation, by interfering with the progression of the cellular cycle and inducing apoptosis by modulation of signal transduction pathways. The cancer is a global public health problem with increasing levels of mortality. Although exists a great variety and types of cancer, we can remark the role of vitamin C and its effects in this suffering. AA is effective protecting against the oxidative damage in tissue and also suppressing the carcinogens formation like nitrosamines.

Although vitamin C is a cytotoxic agent for tumor cells and non toxic for normal cells, in modern medicine and conventional favors more the use of powerful toxic chemotherapeutic agents. A great amount of studies have shown that the consumption of vitamin C is inversely related with cancer with protective effects in cancer of lung, pancreas, stomach, cervix, rectum and oral cavity.

The guanine oxidation, a DNA purine, gets reduced significantly after the vitamin C supplementation, but the adenine oxidation, another purine, it's up high which suggest the antioxidant role of the vitamin. Other extensive studies both *in vivo* and *in vitro* have shown its ability to prevent, reduce or increase the adverse effects of chemotherapy. The combination of vitamin C and vitamin K already given in the chemotherapy increases the survival and the effects of various chemotherapeutic agents in a tumor-ascitic-murine model. The vitamin C has shown be safe even with the radiotherapy. The co administration of vitamin A, β-carotene, E and C can reduce the incidence and delay the progression of several cancers, such as skin, colon, stomach, esophagus, mammary gland and matrix.

Epidemiologic studies have revealed an inverse relation between the consumption of vitamin A, β-carotene, E and C and the incidence of several human cancers. There are a decrease in the risk and incidence of cancer in populations with high content of vitamins in plasma. The carcinogenesis is related with the cell differentiation, progression, metabolism and synthesis of collagen. The basic mechanism for the carcinogenesis is the cell differentiation because the cancer develops when a lost in this differentiation exists. And here is where the mentioned vitamins have a wide influence over de cell growth and its differentiation. Vitamin C is a strong antioxidant that acts synergistically with vitamin E in the purification of free radicals which are carcinogenic. The AA as sodium ascorbate exerts marked cytotoxic effects over many human cell lines when they are cultured. These effects are dose dependent.

Lupulescu reported that vitamin C (up to 200 ug/mL) did not cause any morphological change in mouse melanoma, neuroblastoma, and mouse and rat gliomas but is lethal for neuroblastoma cells. Cytotoxic effects are dependent cell also because they are stronger in human melanoma cells compared to mouse melanoma. It has even been suggested that the cytotoxicity induced by vitamin C is mediated primarily by the formation of hydrogen peroxide in the cell surface. The cytotoxic activity may also be mediated by the presence of cupric ions (Cu^{2+}) in malignant melanoma cells that react with vitamin C to form free radicals in solution. Vitamin C also invests into cells, transforming them chemically to a normal phenotype fine.

Studies of cell surface and ultrastructure suggest that cancer cells after administration of vitamin C had cytolysis, cell membrane damage, mitochondrial changes, nuclear and nucleolar reduction and an increase in the formation of phagolysosomes. Changes in cell surface as cytolysis showed predominantly increased synthesis of collagen and disruption of the cell membrane with increased phagocytic activity and apoptotic.

The quantitative estimation of cellular organelles shown that vitamin C affects the intracellular distribution of the organelles, event that plays an important role in the citodifferentiation of the carcinogenic cell and this is the shared effect that not only vitamin C has, but also vitamin A and E. Changes in the Golgi complex and apoptotic activity and autophagic addition to changes in cell surface and in some cases even the reversal of transformed cells to their normal cell types are needed in the possible reduction in incidence of various cancers.

Have also been associated changes in the protein synthesis, DNA and ARA with the differentiation and proliferation cell. But these mechanisms are not clear yet. It have been mentioned that many of this metabolic effects are mediated by the transcription and translation at genomic level. This vitamins modulate the DNA synthesis and the genetic expression in a similarly to hormones and steroids.Their effects can affect the chemical mutagenicity and the cell status. The vitamins can control the cell replication affecting the DNA, RNA and proteins in specific places which are target of electrophyles, promoting the rearrangement of codons in the altered cells and the translocation of specific genes or carcinogens. In this way, vitamins A, E and C affects directly the DNA, RNA and the protein synthesis in the carcinogen cells.

The vitamin C administration decrease the DNA synthesis in the core, the RNA synthesis in the nucleolus and the protein in the cytoplasm of these cells. This inhibition is accompanied by ultrastructural changes mentioned which decreases the cancer progression.

Mechanism of action:

Have been proposed many mechanisms of the vitamin C activity in the prevention and treatment of cancer:

1. Would improve the immune system by increasing the lymphocytes production.

2. Stimulation in the collagen synthesis.

3. Inhibition of the hyaluronidase, keeping the substances around the tumor intact avoiding metastasis.

4. Inhibition of carcinogen virus

5. Correction of a likely ascorbate deficiency, seen in patients with cancer

6. Adequate healing after the surgery.

7. Improvement in the effect of some chemotherapeutic agents, just like tamoxifen, cisplatina, DTIC and others.

8. Reduction of the toxicity of other chemotherapeutic agents like adriamicine.

9. Prevention of the cell damage by free radicals.

10. Neutralization of carcinogenic substances.

Patients with cancer tend to immune-undertake, showing low levels of ascorbate in their lymphocytes. The survival of immune system is important both for inhibit the carcinogen cell growth phase and to prevent its proliferation. The supplementation with ascorbate increases the number and the effectiveness of the lymphocytes and upgrades the phagocythosis

The characteristics of the neoplastic cell and its behavior (invasiveness, selective nutrition and possibly accelerated growth) are caused by microenvironmental depolymerization. This destabilization of the matrix is favored by constant exposure to lysosomal glycosidases continually released by the neoplastic cell. The AA is then involved in the control and restriction of this degradative enzyme activity.

The synthesis of collagen is a major factor for the encapsulation of tumors or metastases decreased via the development of a nearly impermeable barrier. AA is necessary for synthesis of collagen and its stabilization. A loss of ascorbate significantly reduces the hydroxylation of proline and hydroxyproline and hydroxylysine to lysine respectively, affecting the cross linking of collagen. This disrupts the structure of collagen triple helix, which increases its catabolism s. *In vitro*, vitamin C also increases the synthesis of collagen in fibroblasts.

11. Ascorbic acid in diabetes mellitus

It has been demonstrated *in vitro* competition between glucose and ascorbic acid by the cell membrane transporter, granulocytes and fibroblasts, and under conditions of substantial and significant changes in chemotaxis of PMN leukocytes and mononuclear cells. There are significant changes to various chemoattractants, and a significant correlation with the decrease in AA.

These results are consistent with the hypothesis that the chronic hyperglycemia associated with leukocyte AA deficiency, an acute inflammatory response damaged and altered susceptibility to infections, and failure to repair bleeding in these patients, further changes are observed sustained hyperglycemia.

The concentrations of ascorbic acid (AA) are decreased in tissues and plasma in diabetes. These values can be normalized with extra supplements of 20-40 mg/d or corresponding to its maximum synthetic rate. Treating diabetic rats with this scheme prevents the decreased activity of granulation tissue of proline hydroxylase (Prolasa) an AA-dependent enzyme, required to maintain the normal properties of collagen. The decrease in AA concentration in plasma and Prolasa activity in diabetes can be normalized by the inhibitor of aldose reductase. We conclude that in diabetic animals there is a deficiency of AA, which may be responsible for the observed changes in collagen in diabetes.

The decrease in plasma ascorbic acid in diabetes plays an important role in the abnormalities of collagen and proteoglycans. These are the 2 major constituents of the extracellular matrix and its abnormalities are associated with the pathogenesis and complications of diabetes. The structural similarity between glucose metabolism and AA and can interact at the level of the membrane and transporters.

Ascorbic acid enhances the collagen and proteoglycans in fibroblast culture media. This stimulatory action is inhibited by high concentrations of glucose (25 mM). This effect however, is not mediated AA consumption of fibroblasts. Insulin removes the inhibitory effect of glucose on the production of collagen, but the mechanism is not yet known. Thus high concentrations of glucose in diabetes damage the action of ascorbic acid at the cellular level.

12. Ascorbic acid in essential hypertension (HAS)

High blood pressure is a powerful indicator of heart disease and stroke. And in many cases is "asymptomatic" or people who have it doing not give importance. However there have been great efforts to use its measurement in the detection of primary or secondary essential hypertension for decades.

Virtually the observed declines in blood pressure and its control in recent years due to better control among individuals diagnosed as hypertensive. In this regard dietary factor has been the best for control. Obesity, dietary sodium and alcohol consumption are strongly associated with low or high blood pressure values.

A high intake of polyunsaturated fatty acids and magnesium are associated with for instance with low pressure. It has also shown an inverse association between plasma vitamin C and blood pressure. Subjects with serum levels of vitamin C equal to 0.5 mg/dL have a systolic pressure 122 mm Hg average compared to subjects with average pressure is 113 mm Hg and whose values of serum vitamin C are 0.9 mg/dL, showing a relative difference of 7%.

These subjects have a similar difference in diastolic pressure ranging from 78 to 73 mm Hg, a difference of 6%. The prevalence of hypertension was 7.5% in the group of subjects with low serum vitamin C and only 1% in subjects with high values of the vitamin in the serum. These results were consistent in several studies regardless of quintiles being compared.

Such relationships have also been identified in Chinese-American population; both men and women aged 60-96 years without antihypertensive treatment. This study revealed a statistically significant difference between the values of systolic and diastolic pressure in upper and lower quintiles of 14% (21 mm Hg) and 9% (8mmHg) respectively. It appears that vitamin C has a lowering effect on systolic rather than diastolic pressure. Supplementation with vitamin C (1g/day) does not influence the diastolic pressure. Subjects with low vitamin C levels in serum have a high risk of developing stroke compared with those with high values in plasma of the vitamin. Hypertensive subjects, usually overweight, and low levels of serum vitamin C have the same risk.

The increase in the consumption of vitamin C during periods of fat restriction occurs on the one hand a reduction in blood pressure. Low levels of AA in plasma are also associated with low concentrations of 6-keto-prostaglandin F, a prostacyclin. Thus dietary antioxidants enhance the production of prostacyclin for the purification of free radicals and peroxides that inhibit prostacyclin synthase. Vitamin C and blood pressure then are related, because it has a lowering effect on blood pressure especially when fat intake is low.

13. Ascorbic acid and cardiovascular disease

Vitamin C acts as a regulator of the catabolism of cholesterol into bile acids in the guinea pig and is an important factor in the regulation of lipid in several animal species (rabbit, horse, and rat).

Correlation studies in humans have shown an inverse relationship between vitamin C intake and mortality from cardiovascular disease.

Experimental and observational studies in humans have been inconsistent but indicate that individuals with high cholesterol consumption, greater than or equal to 5.20mM/L (200mg/dL) and lower in tissue saturation, increase concentration of vitamin C, which may have a beneficial effect on total cholesterol. This effect is explained by the promotion or inhibition of degradation of prostacyclin and its implications for thrombosis and atherogenesis, in addition to its protective effect on lipid peroxidation. In patients with high cardiovascular risk, supplementation with antioxidant vitamins shows no reduction in overall mortality or incidence of any vascular disease, cancer or other adverse events.

Recent findings indicate a relationship between the nutritional status of vitamin C (as measured by the concentration of ascorbate in serum), biological markers of infection and haemostatic factors and support the hypothesis that vitamin C may protect against cardiovascular events through effects on the haemostatic factors in response to infection.

This relationship is surprising given the uncertainty and potential error in the estimation of consumption of vitamin and vitamin C status assessment (determined mostly by food intake records of 24 h blood samples isolated). Add to this the wide variation between subjects is greater than within the same subject.

Lower socioeconomic status and smoking are associated with low concentrations of ascorbate and high concentrations of homeostatic factors that may be confounding factors in cross-sectional studies.

As expected smokers have lower concentrations of AA in serum than non-smokers. The relationship between concentrations of ascorbate and homeostatic factors is very consistent, but when cigarette smokers were excluded from the analysis, we obtain an association where smoking becomes a confounding variable. The relationship between fibrinogen, Factor VIIc and ascorbate concentrations were consistent in subjects taking supplements of vitamin C during 1 year, which indicates that the homeostatic factors relate the variability in the status of vitamin C within a range usual dietary.

The inverse association between homeostatic factors and serum concentrations of ascorbate is strong and consistent, however only some markers of infection (e.g. C-reactive protein and α1-antichymotrypsin) are related inversely and significantly with serum ascorbate. It is possible that this low concentration of ascorbate may be the result rather than the cause, of a biological response to infection. The strong relationship between serum ascorbate and dietary intake suggest however that their serum concentrations reflect the nutritional status of the vitamin.

The various studies reported in the literature indicate that vitamin does not prevent respiratory infection but may modulate the biological response, leading to less severe disease, so it has a protective function in lung function.

13.1. Effect of antioxidants in cardiovascular disease

It has been suggested a protective effect of antioxidants such as vitamin C, A (β-carotene) and E plus selenium in cardiovascular disease. Prospective studies so far have documented an inverse relationship between vitamin C intake and cardiovascular disease, and a strong protective effect of vitamin E supplementation on coronary patients.

Finnish and Swiss studies showed that blood levels of ascorbate and therefore a diminished nutritional status of vitamin predicts myocardial infarction. Low levels of vitamin C increased to 2.7 times the risk of myocardial infarction and this is independent of other risk factors. Mediterranean studies showed a 70% reduction in mortality and risk of myocardial infarction independent of the effect on blood pressure and lipids. Any protective effect on heart disease of these antioxidants is mediated by the oxidation of LDL cholesterol, but there may be other mechanisms of homeostasis and inflammation.

13.2. Infection, homeostasis, and cardiovascular disease

Fibrinogen and factor VII are recognized risk factors of myocardial infarction and stroke, in the same way that acute and chronic infection and increased white blood cell count are risk factors for cardiovascular disease. The infection may contribute to the inflammatory process observed in atherosclerosis.

C-reactive protein and alpha-1 antichymotrypsin are acute phase proteins are synthesized in hepatocytes in large numbers in inflammatory processes. This synthesis is mediated primarily by IL-6 produced by monocytes and macrophages. Elevated fibrinogen favors these mechanisms and therefore an increased cardiovascular risk. In this way a reduction in dietary intake in winter for instance, would lead to lower serum ascorbate levels, an increase in susceptibility to infection and the factors haemostatic factors and therefore to an increase in cardiovascular mortality.

According to the seasonal variations of vitamin C intake may be relatively low (<80 mg/day on average), corresponding to serum ascorbate concentrations of 50 umol/L. within this range may be variations in infection and homeostatic markers. Increased intake of vitamin C to 90-100 mg/day can increase in these subjects more than 60 umol/L, which has a significant effect on all risk factors.

14. Ascorbic acid and immunity

In stress situations the adrenal glands react liberating a large number of active and ready hormones. It has been suggested that 200 mg of vitamin C per day can reduce stress levels caused by these hormones. The stress suppresses the immune response. Megadoses of vitamin C increases the body levels of antibodies in animal models (rats stressed and unstressed) having the highest values stressed rats. Stressed animals may need more vitamin C for proper immune system function.

Healing is characterized by synthesis of connective tissue, whose main component is collagen. This molecule AA required for cross linking of the fibers in hydroxylated residues of prolyne and lysine. Ascorbic acid supplementation is necessary for healing since this is oxidized during the synthesis of collagen.

There is an undeniable evidence of the interaction of vitamin C and phagocytes. The collected cells from the blood, peritoneal or alveolar fluid usually contain high concentrations of vitamin C (1-2 ug/mg protein). Guinea pig neutrophils produced H_2O_2 and destroy staphylococci in the same way they do control cells. Both ascorbate as dehydroascorbic acid are used for phagocytic process. Neutrophils can avoid self-poisoning absorb extra amounts of ascorbic acid, which can neutralize the antioxidants. However glucolytic activity does not increase much in the neutrophils of guinea pigs supplemented and the stimulation of NADPH oxidase activity is depressed. The addition of ascorbate to the culture media of normal macrophages increases the concentration of cyclic GMP (cGMP) in addition to the route of pentose or hexose monophosphate.

Although the addition of large amounts of ascorbate can inhibit myeloperoxidase activity is not altered its bactericidal capacity. It has been an increase in the bactericidal activity in mouse peritoneal macrophages by the addition of ascorbate to the medium. Besides ascorbate increase the motility and chemotactic activity of these cells. The motor functions of cells as the random motion and chemotactic migration of neutrophils and macrophages is damaged in the absence of vitamin C. Ascorbic acid can also influence the ability of certain cell lines to produce interferon. The addition of AA to cultures of skin embryonic or fibroblasts leads to the production of interferon.

Vitamin C is also necessary for thymic function and operation of certain cells involved in the production of thymic humoral factor. Thymic content of dehydroascorbate diminishes in direct proportion to vitamin C intake. The hormonal activity of thymic extracts correlates with thymic ascorbate and inversely with dehydroascorbate.

15. Ascorbic acid and gallbladder

The gallbladder disease is highly prevalent in the U.S. and in Mexico, remains a serious public health problem. It has been estimated that only about in U.S. 20 million Americans have gallstones partially or entirely composed of cholesterol. Gallstones form when bile supersaturated with cholesterol is destabilized. AA affects a limiting step in the catabolism of cholesterol into bile acids in experimental animals, as described AA-deficient guinea pig common development of cholesterol gallstones.

Because of this it has been hypothesized that the deficiency in humans may be a risk factor for this disease in humans. The main findings in humans have shown an inverse relationship between serum levels of AA and biliary disease prevalence among women. It was also observed a low prevalence of clinical biliary disease between women taking ascorbic acid supplements. However, this prevalence has not been observed in males. Almost all the findings from different countries agree to respect.

In another study, Simon showed that the use of ascorbic acid supplementation correlates with biliary disease among postmenopausal women with coronary disease. Among women who consumed alcohol, the use of ascorbic acid supplementation was associated independently with a 50% reduction in the prevalence of gallstones and 62% for cholecystectomies.

Within the NHANES III study was not observed association, linear or nonlinear, between serum ascorbic acid and prevalence of biliary disease in men. Reflecting the low prevalence of the disease in men and reduced statistical power to detect such an association.

It has been hypothesized that the inverse relationship between AA and biliary disease, demonstrated in animals, affects the activity of the enzyme cholesterol-7-α-hydroxylase, which is the limiting step that regulates the metabolism of cholesterol into bile acids. Supplementation with ascorbic acid increases the activity of the enzyme up to 15 times compared with the vitamin-deficient animals that develop the formation of cholesterol gallstones. Additionally there is a hypersecretion of mucin, a glycoprotein that is secreted by the epithelium of the gallbladder, which precedes cholesterol destabilization and gallstone formation. Because oxygen and hydroxyl radicals stimulate mucin hypersecretion, inhibition of the oxidative changes in the vesicle due to AA can decrease the production of this glycoprotein.

16. Ascorbic acid in other conditions

16.1. Sjögren's syndrome

16.1.1. Vitamin C and Sjögren syndrome

Primary Sjögren's syndrome (SSP) is a chronic disorder of unknown cause, characterized by dry eyes and mouth. It usually occurs in middle-aged women with a prevalence of 1:5000. Patients may have swelling of joints, muscles, nerves, thyroid, kidneys or other body areas. These symptoms result from lymphocytic infiltration and destruction of these tissues.

The diagnosis is based on clinical examination of the eyes and mouth, blood tests specific (auto antibodies) and biopsy of minor salivary gland (taken from inside the inner lip). Sjögren's syndrome is not fatal, but must be addressed quickly to prevent complications due to dry mouth (caries, abscesses, gingivitis) and eyes (corneal erosion, infection). However, there is no therapy available that "removes" these symptoms because all therapies are directed at eliminating the symptoms and prevent complications.

On the other hand there are several studies that reported a direct relationship between the clinical manifestations of Sjögren's syndrome and those caused by deficiency of ascorbic acid (vitamin C): scurvy. A frank deficiency of vitamin C causes scurvy, a disease characterized by multiple hemorrhages. Scurvy in adults is manifested by latitude, weakness, irritability, vague muscle pain, joint pain and weight loss. Early signs are objective as bleeding gums, tooth loss and gingivitis.

The diagnosis of scurvy, is achieved by testing plasma ascorbic acid, low concentration indicates low levels in tissues. It is generally accepted that ascorbic acid concentration in the layer of coagulated lymph (20-53 ug/10^8 leukocytes) is the most reliable indicator of nutritional status regarding vitamin C and its concentration in tissues and serum.

16.2. Pharmacological data

Ascorbic acid is specific in the treatment of scurvy; the dose required can best be measured by determining urinary excretion after a dose of saturation, depending on the speed at which the saturation is required is the recommended daily dose ranging from 0.2 and 2.0 g/day. In the vitamin deficiency C tissue saturation is achieved with three daily doses of 700 mg c/u, for three days. Harris defined as saturation of tissues, a sufficient store where an ascorbic acid excretion 50 mg or more occurs in a period of 4 to 5 hours after a dose of 700 mg/day.

16.3. Previous experience in animals

Kessler in his study reported that rats which have been induced vitamin C deficiency, develop various manifestations of primary Sjögren's syndrome (SSP), such as infiltration of mononuclear cells in salivary and lachrymal glands and that these were more severe is female rats than in male rats, concluding that these pathological changes are similar to those that characterize the syndrome in humans.

In previous studies Hood reported a direct relationship between the manifestations of primary Sjögren's syndrome (xerostomia and xerophthalmia) and clinical signs of scurvy such as gingivitis, periodontal bleeding and loss of teeth. In their study, Hood study 5 subjects men whose diets did not contain ascorbic acid, for 84 to 97 days. During the deficiency, in the demonstrations that make Sjögren syndrome, observed that prostaglandins, particularly PGE_1, is important in the immune defects associated with the syndrome.

In 1992, Gomez et al, from the National Institute of Medical Sciences and Nutrition "Salvador Zubiran", observed values less than 0.2 mg AA/dl in plasma of patients with SSP (reference values were from 0.4 to 2.0 mg/dL), representing a frank deficiency of vitamin in 100% of cases with SSP.

16.4. Role of vitamin C in other body disorders

It is reported that the diabetic individual has low levels of vitamin C in plasma and leukocytes, which is our immune defense. However, more clinical studies, in a large scale, are needed to determine whether the supplementation with large doses of the vitamin are beneficial or not. Some studies have shown that supplementation with 2 g/d, decreased glucose levels in diabetics and reduce capillary fragility.

Megadoses of vitamin C can still be toxic in diabetics with kidney disorders. It was mentioned that vitamin C also helps to reduce body glycosylation, which shows abnormalities in the binding of sugars and proteins. In addition vitamin C reduces the accumulation of sugar sorbitol which damages eyes and kidneys. Vitamin C lowers blood pressure and plasma cholesterol helping to keep the blood flowing and protected from oxidation in a synergistic action with vitamin E. In doses of 1g/day protects the body against LDL lipoproteins.

Atherosclerosis is the best contributor to heart disease. Vitamin C prevents the formation of atheromatous plaque by inhibiting the oxidative modification of LDL's, which contributes to atherosclerotic process for their cytotoxic effects, inhibition of receptor radical scavengers and their influence on the motility of monocytes and macrophages.

Vitamin C also helps to prevent atherosclerosis through the synthesis of collagen in the arterial wall and prevent undesirable adhesion of leukocytes to the damaged artery.

Supplementation with 2 g/day reduces the adhesion of monocytes to blood vessels, effectively reverses the vasomotor dysfunction observed in patients with atherosclerosis. In addition these doses increase HDL, being highly protective against heart attacks and stroke. Risk is reduced by up to 62% in subjects consuming 700 mg/day compared with those consuming 60mg/day or less. Only Joel study has shown that low levels of ascorbic acid in serum (AAs) are marginally associated with an increased risk and fatal cardiovascular disease was significantly associated with an increased risk of mortality for all causes. Low levels of AAs are also a risk factor for cancer death in men, but unexpectedly it was associated with a decreased risk of cancer death in women.

Vitamin C has an effect antihistaminic. Subjects with low plasma vitamin C levels have elevated blood histamine and vitamin supplementation, reduces these levels. Table 7 shows the relationship of vitamin C reported in different conditions.

Low concentrations of AA	High concentration of AA
Rheumatoid arthritis (Lunec)	Cancer in women (Joel)
Cancer in Men (Joel)	Uric Acid Excretion (Stein)
Asthma (Ruskin)	Back pain and spinal discs.
Bronchospasm	Antioxidant (Kahn)
Cataract (Jackes)	Allergic process (Ruskin)
Aging (Jackes)	Blood pressure (Ringsdorff)
Retinopathy (Crary)	Constipation (Sindair)
(Macular Degeneration)	Probable association with menopause. (Smith)

Figures in parentheses indicate the reference.

Table 7. Association of serum vitamin C (ascorbic acid) in serum or plasma in different symptoms and diseases.

Author details

José Luis Silencio Barrita[1] and María del Socorro Santiago Sánchez[2]

1 Association of Chemists, National Institute for Medical Sciences and Nutrition "Salvador Zubiran", Mexico

2 Department of Nutrition and Dietetics, General Hospital No, 30 "Iztacalco", Mexico

References

[1] Akhilender Naidu K Vitamin C in human health and disease is still a mystery? An overview. Nutr J 2003, 2:1475-2891-2-7

[2] Anitra C Carr and Balz Frei.Toward a new recommended dietary allowance for vitamin C based on antioxidant and health effects in humans Am J Clin Nutr 1999;69:1086–107

[3] Ascencio C., Gomez E., Ramirez A., Pasquetti A., Bourges H. Dietary habits and nutrient intake in patients with primary Sjögren's syndrome.7th International Congress of Mucosal Immunology. August 16-20, 1992, Prague, Czechoslovaquia.

[4] Bergman F, Curstedt T, Eriksson H, van der Linden W, Sjo°vall J. Gallstoneformation in guinea pigs under different dietary conditions: effect of vitamin C on bile acid pattern. Med Biol. 1981; 59:92-98.

[5] Blomhoff R. Dietary antioxidants and cardiovascular disease Curr Opin Lipidol 16:47–54. # 2005

[6] Bowes and Church.- Food values of portions commonly used. Lippincot Co.,1975.

[7] Bourges H. Madrigal H. Chavéz A., Tablas del valor nutritivo de los alimentosmexicanos. Publicación L-12 INN, 1983, 13 ed. México.

[8] Bradley A.V. Tables of food values, Chas A. Bennet Co. 1976.

[9] Cameron E, Pauling L. Supplemental ascorbate in the supportive treatment of cancer: Prolongation of survival times in terminal human cancer. Proc Natl Acad Sci 1976; 73:3685-3689.

[10] Chen Qi, Espey M G, Krishna MC, Mitchell JB, Corpe CP, Buettner GR, Shacter E, Levine M. Pharmacologic ascorbic acid concentrations selectively kill cancer cells: Action as a pro-drug to deliver hydrogen peroxide to tissues. PNAS 2005, 102(38): 13604-13609

[11] Crary, E.J. and M.F. McCarty, 1984. Potential clinical applications for high-dose nutritional antioxidants. Med. Hypoth., 13: 77-98.

[12] Drake IM, Davies MJ, Mapstone NP, Dixon MF, Schorah CJ, White KLM, Chalmers DM, Axon ATR. Ascorbic acid may protect against human gastric cancer by scavenging mucosal oxygen radicals. Caranogenesis. 1996, 17 (3):.559-562,

[13] Engelfried J.J. The ascorbic acid saturation test. J.Lab. Med. 1944, 29:234

[14] F.J. de Abajo y M. Madurga. Vitamina C: aplicaciones terapéuticas en la actualidad. Med Clin (Barc) 1993; 101: 653-656

[15] Fisher E, McLennan SV, Tada H, Heffernan S, Yue DK, Turtle JR. (Interaction of ascorbic acid and glucose on production of collagen and proteoglycan by fibroblasts. Diabetes, 1991, 40(3):371-6)

[16] Gueguen S, Pirollet P, Leroy P, Guilland JC, Arnaud J, Paille F, Siest G, Visvikis S, Hercberg S, Herbeth B. Changes in Serum Retinol, α-Tocopherol, Vitamin C, Carotenoids, Zinc and Selenium after Micronutrient Supplementation during Alcohol Rehabilitation. J Am Coll Nutr, 2003, 22:303-310

[17] Gomez E, Silencio JL, Bourges H. Vitamin B2, B6 and C status in patients with primary sjögren's syndrome. 7th International congress of mucosal immunology, august 16-20, 1992, Prague Csechoslovaquia.

[18] Harris J.L. Vitamin C saturation test. Standarization meassurements at graded levels of intake. Lancet 1943, 1:515,

[19] Head KA, Ascorbic Acid in the Prevention and Treatment of Cancer. Altern Med Rev 1998; 3(3):174-186)

[20] Herrera, V. Más evidencia en contra del uso de vitaminas y antioxidantes en la prevención de enfermedades crónicas Evidencia Actualización en la Práctica Ambulatoria- 2002, 5(6): Nov-Dic

[21] Hoffman-La Roche F. Compendio de vitaminas, 1970, Basilea, Suiza.

[22] Hood J., Burns Ch.A., Hodges R.E. Sjögren's syndrome in scurvy. N. Engl. J. Med. 1970, 282:1120

[23] Horrobin D.F., Manku M.S. The regulation of prostaglandin E1 formation: A candidate for one of the fundamental mechanism involved in the actions of vitamin C. Med. Hypotheses. 1979, 5:849.

[24] Ip C. Interaction of vitamin C and selenium supplementation in the modification of mammary carcinogenesis in rats. J Natl Cancer Inst 1986;77:299-303.

[25] Iqbal K, Khan A Ali Khan Khattak MM. Biological Significance of Ascorbic Acid (Vitamin C) in Human Health: A Review Pakistan J Nutr 2004, 3(1):5-13,

[26] Jacob RA, Pianalto FS, Agee RE. Cellular ascorbate depletion in healthy men. J Nutr. 1992; 122:1111-1118.

[27] Jacques, P.F., L.T. Chylack, R.B. McGandy and S.C Hartz, 1988. Antioxidant status in persons with and without senile cataract. Arch Ophthalmol, 106: 337-340.

[28] Jenkins SA. Biliary lipids, bile acids and gallstone formation in hypovitaminotic C guinea-pigs. Br J Nutr. 1978;40:317-322

[29] Jenkins SA. Hypovitaminosis C and cholelithiasis in guinea pigs. Biochem Biophys Res Commun. 1977; 77:1030-1035.

[30] Kahn, H.A., H.M. Leibowitz and J.P.Ganley, 1977. The Framingham Eye Study: Outline and major prevalence findings. Am. J. Epidemiol., 106: 17-32.

[31] Kathleen A. Head, N.D. Ascorbic Acid in the Prevention and Treatment of Cancer Altern Med Rev 1998; 3(3):174-186

[32] Kessler S. A laboratory model for Sjögren syndrome. Lancet 1968, 52:671

[33] Khaw KT, Woodhouse P. Interrelation of vitamin C, infection, haemostatic factors, and cardiovascular disease BMJ 1995; 310:1559-1563

[34] Lunec, J.B., 1985. The determination of dehydroascorbic acid and ascorbic acid in the serum and sinovial fluid of patients with rheumatoid arthritis. Free Radical Research Communications, 1: 31-39.

[35] Lupulescu A The Role of Vitamins A, B Carotene, E and C in Cancer Cell Biology. Intern. Vit. Nutr. Res. 1993; 63:3-14.

[36] Marks J. A guide to the vitamins, their role in health and disease. MTP Medical and Technical Pub. 1975, England.

[37] McLennan S, Yue DK, Fisher E, Capogreco C, Heffernan S, Ross GR, Turtle JR. Deficiency of ascorbic acid in experimental diabetes. Relationship with collagen and polyol pathway abnormalities. Diabetes, 1988 ,37(3):359-361,

[38] MRC/BHF Heart Protection Study of antioxidant vitamin supplementation in 20.536 high-risk individuals: a randomised-placebo controlled trial. Heart Protection Study Collaborative Group. The Lancet 2002;360:23-33

[39] Mullan BA, Young IS, Fee Hd, McCance DR. Ascorbic Acid Reduces Blood Pressure and Arterial Stiffness in Type 2 Diabetes Hypertension. 2002; 40:804-809

[40] Omaye, T.,Scala J.H., Jacob R.A. Plasma ascorbic acid in adult males: effect of depletion and supplementation. Am. J. Clin. Nutr. 1986, 44: 257,

[41] Paul A.A., Southgate D.A. The composition of foods, 14th edition Elsevier/North Holland Biomedical Press, 1985.

[42] Pecoraro RE, Chen MS. Ascorbic acid metabolism in diabetes mellitus. Ann New York Acad Sci, 1987, 498(1): 248-258,

[43] Ringsdorf, W.M. Jr. and E. Cheraskin, 1981. Ascorbic acid and glaucoma: A Rev. J. Holistic. Med., 3: 167-172.

[44] Roe J.H., Kuether C.A. determination of ascorbic acid in wholeblood and urine through the 2,4-dinitrophenylhydrazine derivative of dehydroascorbic acid. J. Biol. Chem, 1942, 147:399 .

[45] Rogur L, Lundblad and Fiona M . MacDonald. Handbook of Biochemistry and Molecular Biology, Fourth Edition, Edited by CRC Press 2010 Pages 243-250

[46] Rose C. R., Nohrwld L.D. Quantitative analysis of ascorbic acid and dehydroascorbic acid by High Performance Liquid Chromatography Anal. Biochem. 114:140, 1981.

[47] Rune Blomhoff. Dietary antioxidants and cardiovascular disease. Curr Opin Lipidol 16:47–54. # 2005

[48] Ruskin, S.L., 1947. Sodium ascorbate in the treatment of allergic disturbances. The role of adrenal cortical hormone-sodium-vitamin C. Am. J. Dig. Dis.,14: 302-306

[49] Sahyoun, R.N.,1996. Carotenoids, vitamins C and E and mortality in an elderly population. Am. J. epidemio.,144: 501-511.

[50] Sánchez-Moreno C, Cano MP, de Ancos B, Plaza L, Olmedilla B, Granado F, Martín A.Effect of orange juice intake on vitamin C concentrationsand biomarkers of antioxidant status in humansAm J Clin Nutr 2003; 78:454–60.

[51] Silencio Barrita JL Vitaminas: conceptos generales, Nutr clin 2006;9(3):36-44

[52] Simon JA., Esther SH, Jeffrey AT. Relation of serum ascorbic acid to mortality among US adults. J. Am. College of Nutr, 2001.20(3);456

[53] Simon JA, Hudes ES. Serum Ascorbic Acid and Gallbladder Disease Prevalence Among US Adults. The Third National Health and Nutrition Examination Survey (NHANES III) Arch Intern Med. 2000; 160:931-936

[54] Simon JA. Ascorbic acid and cholesterol gallstones. Med Hypotheses. 1993; 40:81-84.

[55] Simon J.A Vitamin C and cardiovascular disease: a review J Ame Collage of Nutr, 1992, 11 (2):107-125

[56] Simon JA, Hudes ES, Tice JA. Relation of serum ascorbic acid to mortality among US adults. J Am Coll Nutr 2001; 20:255–63.

[57] Simopoulos AP. Human requirement for N-3 polyunsaturated fatty acids. Poult Sci. 2000 Jul;79(7):961-70.

[58] Spittle C.R. Artherosclerosis and vitamin C. Lancet, december 11:1280, 1971.

[59] Stein, H.B., 1976. Ascorbic acid-induced uricosuria: a consequence of megavitamin therapy. Ann. Intern.Med., 84: 385-388.

[60] Sinclair AJ, Girling AJ, Gray L. an investigation of the relationshipbetween free radical activity and vitamin C metabolism in ederly diabetic subjects with retinopathy. Gerontology 1992, 38:266-274

[61] Smith CJ. Non-hormonal control of vaso-motor flushing in menopausal patients . Chicago med, 1984.

[62] Valk EE, Hornstra G Relationship between vitamin E requirement and polyunsaturated fatty acid intake in man: a review. Int J Vitam Nutr Res. 2000 Mar;70(2):31-42.

[63] Walingo KM. Role of vitamin C (ascorbic acid) on human health- A review. African J Food Agric Nutr Dev (AJFAND): 2005, 5(1):1-25

[64] Watt B.K. A.L. Composition of foods. Agriculture Handbook num. 9, United States, Departament of Agriculture, USA, 1975.

[65] Yi Li and Herb E. Schellhorn*New Developments and Novel Therapeutic Perspectives for Vitamin C J. Nutr. 2007, 137: 2171–2184,

Chapter 5

Food Phenolic Compounds: Main Classes, Sources and Their Antioxidant Power

Maria de Lourdes Reis Giada

Additional information is available at the end of the chapter

1. Introduction

The natural phenolic compounds have received increasing interest in the last years, since a great amount of them can be found in plants and consumption of vegetables and beverages with a high level of such compounds may reduce the risk of development of several diseases due to their antioxidant power, among other factors.

It is known that the metabolism of plants is divided in primary and secondary. The substances that are common to living things and essential to cells maintenance (lipids, proteins, carbohydrates, and nucleic acids) are originated from the primary metabolism. On the other hand, substances originated from several biosynthetic pathways and that are restricted to determined groups of organisms are results of the secondary metabolism [1]. Phenolic compounds are constituted in one of the biggest and widely distributed groups of secondary metabolites in plants [2].

Figure 1 shows the inter-relationships between the primary and secondary metabolism in plants.

Biogenetically, phenolic compounds proceed of two metabolic pathways: the shikimic acid pathway where, mainly, phenylpropanoids are formed and the acetic acid pathway, in which the main products are the simple phenol [3]. Most plants phenolic compounds are synthesized through the phenylpropanoid pathway [4]. The combination of both pathways leads to the formation of flavonoids, the most plentiful group of phenolic compounds in nature [3].

Additionally, through the biosynthetic pathways to the flavonoids synthesis, among the not well elucidated condensation and polymerization phases, the condensed tannins or non-hydrolysable tannins are formed. Hydrolysable tannins are derivatives of gallic acid or hexahydroxydiphenic acid [5].

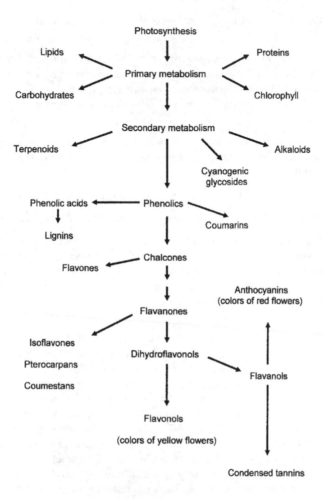

Figure 1. Inter-relationships between the primary and secondary metabolism in plants.

Therefore, phenolic compounds have, as a common characteristic, the presence of at least one aromatic ring hydroxyl-substituted [6]. Another characteristic of these substances is that they are presented commonly bound to other molecules, frequently to sugars (glycosyl residue) and proteins. The existence of phenolic compounds in free form also occurs in plant tissues. However, it is less common, possibly because they are toxic when present in the free state and detoxified, at least in part, when bound.

As a result, phenolic compounds play a role of protection against insects and other animals to the plants. The different types of bond between the glycosyl residue and the flavonoids, such as anthocyanin, also lead to the different derivatives that add colors and color grada-tion to flowers [7].

This way, phenolic compounds are essential to the physiology and cellular metabolism. They are involved in many functions in plants, such as sensorial properties (color, aroma, taste and astringency), structure, pollination, resistance to pests and predators, germinative processes of seed after harvesting and growth as well as development and reproduction, among others [8].

Phenolic compounds can be classified in different ways because they are constituted in a large number of heterogeneous structures that range from simple molecules to highly poly-merized compounds.

According to their carbon chain, phenolic compounds can be divided into 16 major classes [9].

The main classes of phenolic compounds regarding to their carbon chain can be seen in Figure 2.

On the other hand, as to their distribution in nature, phenolic compounds can be divided into three classes: *shortly distributed* (as simple phenols, pyrocatechol, hydroquinone, resorci-nol, Aldehydes derived from benzoic acids that are components of essential oils, such as va-nillin), *widely distributed* (divided in flavonoids and their derivatives, coumarins and phenolic acids, such as benzoic and cinnamic acid and their derivatives) and *polymers* (tan-nin and lignin) [10].

Finally, as to the location in the plant (free in the soluble fraction of cell or bound to com-pounds of cell wall), together with the chemical structure of these substances, phenolic compounds may also be classified as: *soluble* (such as simple phenol, flavonoids and tan-nins of low and medium molecular weight not bound to membranes compounds) and *insoluble* (essentially constituted by condensed tannins, phenolic acids and other phenolic compounds of low molecular weight bound to cell wall polysaccharides or proteins form-ing insoluble stable complexes). This classification is useful from the nutritional view-point, to the extent that the metabolic fate in the gastrointestinal tract and the physiological effects of each group will depend largely on their solubility characteristics. Insoluble phe-nolic compounds are not digested and may be partially or fully recovered quantitatively in the feces, while a part of the soluble can cross the intestinal barrier and be found in the blood, unchanged or as metabolites [3].

The antioxidant activity of food phenolic compounds is of nutritional interest, since it has been associated with the potentiation of the promoting effects of human health through the prevention of several diseases [11]. Additionally, in some cases, these compounds may also be used with therapeutic purposes due to their pharmacological properties [12]. Many phe-nolic compounds with low molecular weight, such as thymol, are used in medicine as anti-septic due to its toxicity [7].

However, the antioxidant activity of phenolic compounds depends largely on the chemi-cal structure of these substances [2]. Among the phenolic compounds with known antiox-idant activity, flavonoids, tannins chalcones and coumarins as well as phenolic acids are highlighted.

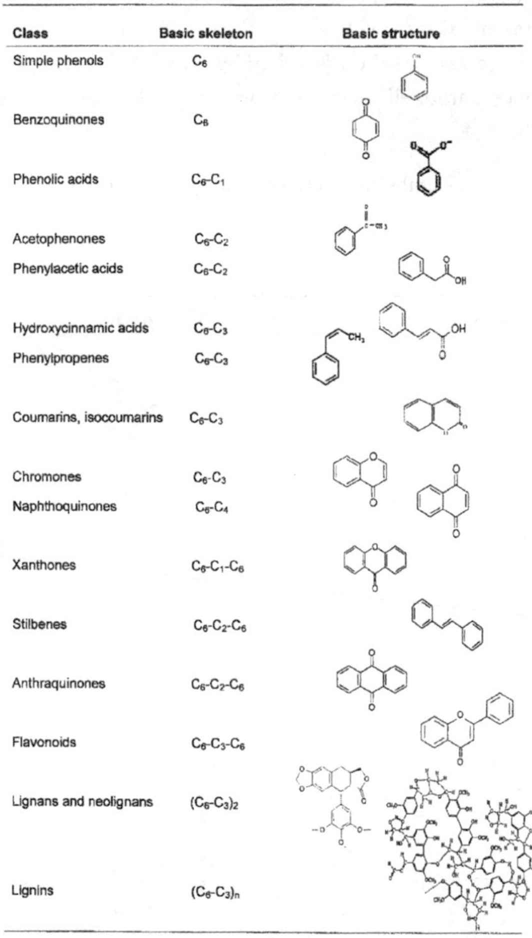

Class	Basic skeleton	Basic structure
Simple phenols	C_6	
Benzoquinones	C_6	
Phenolic acids	C_6-C_1	
Acetophenones	C_6-C_2	
Phenylacetic acids	C_6-C_2	
Hydroxycinnamic acids	C_6-C_3	
Phenylpropenes	C_6-C_3	
Coumarins, isocoumarins	C_6-C_3	
Chromones	C_6-C_3	
Naphthoquinones	C_6-C_4	
Xanthones	$C_6-C_1-C_6$	
Stilbenes	$C_6-C_2-C_6$	
Anthraquinones	$C_6-C_2-C_6$	
Flavonoids	$C_6-C_3-C_6$	
Lignans and neolignans	$(C_6-C_3)_2$	
Lignins	$(C_6-C_3)_n$	

Figure 2. Main classes of phenolic compounds regarding to their carbon chain.

2. Main Classes

2.1. Flavonoids

According to the degree of hydroxylation and the presence of a C_2-C_3 double bond in the heterocycling pyrone ring, flavonoids can be divided into 13 classes [3], the most important being represented by the flavonols, flavanols, flavones, isoflavones, anthocyanidins or anthocyanins and flavanones [2]. Within these classes there are many structural variations according to the degree of hydrogenation and hydroxylation of the three ring systems of these compounds. Flavonoids also occur as sulfated and methylated derivatives, conjugated with monosaccharides and disaccharides and forming complexes with oligosaccharides, lipids, amines, carboxylic acids and organic acids, being known approximately 8000 compounds [13].

The basic chemical structures of the main classes of flavonoids are presented in Figure 3.

Flavonoid	Basic structure
Flavones	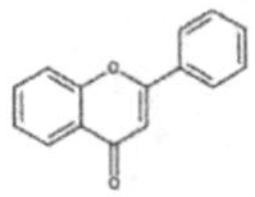
Flavonols	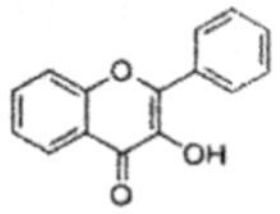
Flavanones	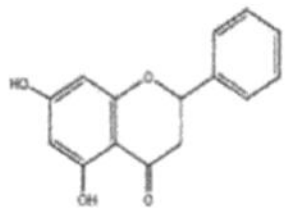
Flavanols	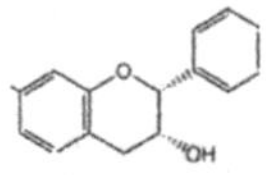
Anthocyanidins	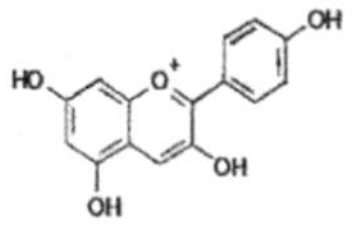
Isoflavones	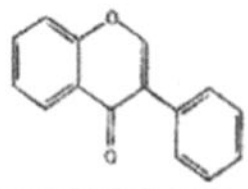

Figure 3. Chemical structures of the main classes of flavonoids.

While members of certain classes of flavonoids (eg., flavonones) are colorless, the other (eg, anthocyanins) are always colored, such as flowers pigments and other plant parts [7].

Flavonoids are important constituents of the human diet [14, 15], and are the most widely distributed phenolic compounds in plant foods and also the most studied ones [10].

It is known that flavonoids are among the most potent antioxidants from plants. The excellent antioxidant activity of these substances is related to the presence of hydroxyl groups in positions 3' and 4' of the B ring, which confer high stability to the formed radical by participating in the displacement of the electron, and a double bond between carbons C_2 and C_3 of the ring C together with the carbonyl group at the C_4 position, which makes the displacement of an electron possible from the ring B. Additionally, free hydroxyl groups in position 3 of ring C and in position 5 of ring A, together with the carbonyl group in position 4, are also important for the antioxidant activity of these compounds [16]. However, the effectiveness of the flavonoids decreases with the substitution of hydroxyl groups for sugars, being the glycosides less antioxidants than their corresponding aglycons [17].

2.2. Tannins

Tannins are phenolic compounds of molecular weight from intermediate to high (500-3000 D) [3] and can be classified into two major groups: hydrolysable tannins and non-hydrolysable or condensed tannins [18]. There is a third group of tannins, phlorotannins, which are only found in brown seaweeds and are not commonly consumed by humans [19].

The hydrolysable tannins have a center of glucose or a polyhydric alcohol partially or completely esterified with gallic acid or hexahydroxydiphenic acid, forming gallotannin and ellagitannins, respectively [20]. These metabolites are readily hydrolyzed with acids, bases or enzymes. However, they may also be oxidatively condensed to other galoil and hexahydroxydiphenic molecules and form polymers of high molecular weight. The best known hydrolysable tannin is the tannic acid, which is a gallotannin consisting of a pentagalloyl glucose molecule that can additionally be esterified with another five units of gallic acid [10].

The condensed tannins are polymers of catechin and/or leucoanthocyanidin, not readily hydrolyzed by acid treatment, and constitute the main phenolic fraction responsible for the characteristics of astringency of the vegetables. Although the term condensed tannins is still widely used, the chemically more descriptive term "proanthocyanidins" has gained more acceptance. These substances are polymeric flavonoids that form the anthocyanidins pigments. The proanthocyanidins most widely studied are based on flavan-3-ols (-)-epicatechin and (+)-catechin [5].

The chemical structures of casuarictin (hydrolysable tannin) and proanthocyanidins (non-hydrolysable or condensed tannins) are shown in Figure 4 A and 4B, respectively.

Although the antioxidant activity of tannins has been much less marked than the activity of flavonoids, recent researches have shown that the degree of polymerization of these substances is related to their antioxidant activity. In condensed tannins and hydrolysable (ellagitannins) of high molecular weight, this activity can be up to fifteen to thirty times superior to those attributed to simple phenols [16].

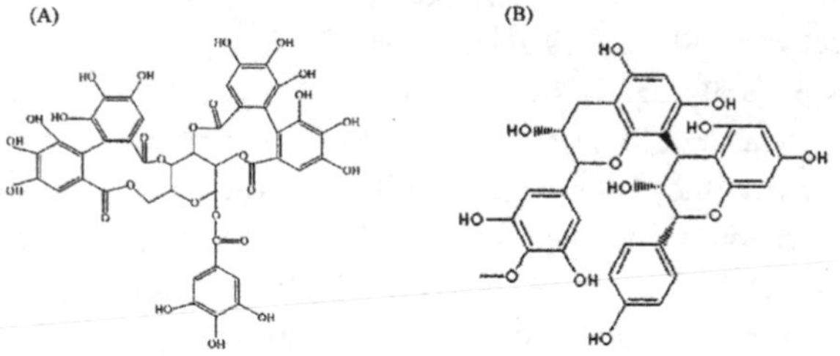

(A)   (B)

Casuarictin (ellagitannin)   Proanthocyanidins

Figure 4. Chemical structures of hydrolysable tannin (A) and non-hydrolysable or condensed tannins (B).

2.3. Chalcones and Coumarins

The chalcones are intermediate in the biosynthesis of flavonoids, being the phloretin and its glucoside phloridzin (phloretin 2'-*o*-glucose), as well as the chalconaringenin and the arbutin, the most frequently found in foods. The phloretin and phloridzin are characteristics of apples, as well as the chalconaringenin is characteristic of tomatoes and arbutin of pears. However, arbutin is also found in strawberries, wheat and its derivatives, as well as in trace amounts in tea, coffee, red wine and broccoli. In some species of plants, the main pigments of yellow flowers are chalcones [21].

Figure 5 shows the chemical estructures of the main chalcones.

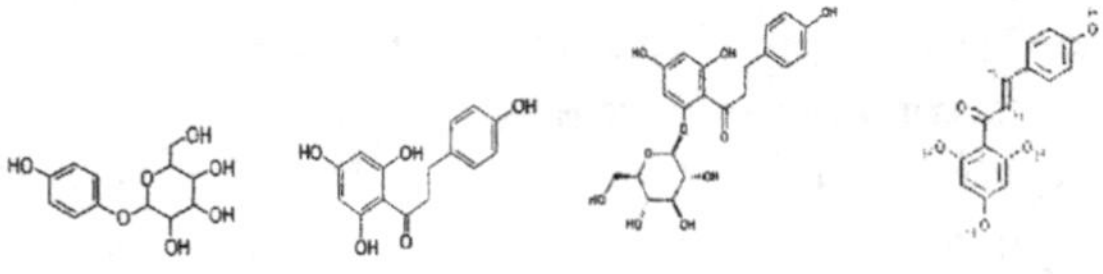

Arbutin   Phloretin   Phloridzin   Chalconaringenin

Figure 5. Chemical structures of the main chalcones.

Like the other phenylpropanoids, coumarins constitute a class of secondary metabolites of plants derivatives from cinnamic acid by cyclization of the side chain of the *o*-coumaric acid [22]. These substances are more common in nature in the form of glycosides, such as umbelliferone, esculetin and scopoletin, and are mainly found in olive oil, oats and spices [3].

The chemical structures of the main coumarins can be seen in Figure 6.

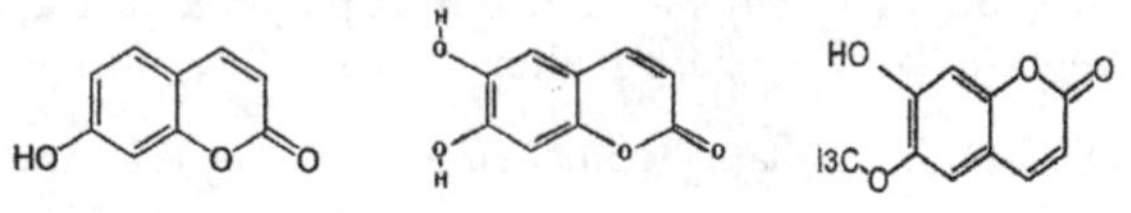

umbelliferone   esculetin   scopoletin

Figure 6. Chemical structures of the main coumarins.

Although the data are still limited, it is known that chalcones and coumarins have antioxidant activity [23].

2.4. Phenolic acids

Phenolic acids can be divided into two groups: benzoic acids and cinnamic acids and derivatives thereof. The benzoic acids have seven carbon atoms (C_6-C_1) and are the simplest phenolic acids found in nature. Cinnamic acids have nine carbon atoms (C_6-C_3), but the most commonly found in vegetables are with seven. These substances are characterized by having a benzenic ring, a carboxylic group and one or more hydroxyl and/or methoxyl groups in the molecule [24].

The general formulas and names of the main benzoic and cinnamic acids are found in Figures 7 and 8, respectively.

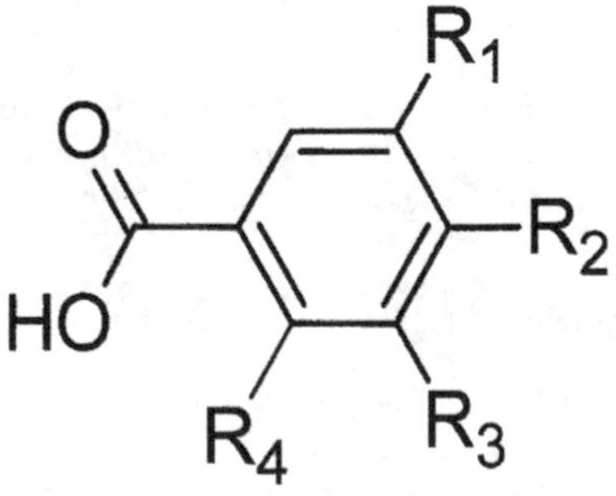

Salicylic acid (R_4 = OH, R_1, R_2, R_3 = H);
Gentisic acid (R_1, R_3 = OH; R_2, R_4 = H);
p-Hydroxybenzoic acid (R_2 = OH, R_1, R_3, R_4 = H);
Protocatechuic acid (R_1, R_2 = OH; R_3, R_4 = H);
Vanillic acid (R_1 = OCH$_3$, R_2 = OH; R_3, R_4 = H);
Gallic acid (R_1, R_2, R_3 = OH; R_4 = H);
Syringic acid (R_1, R_3 = OCH$_3$; R_2 = OH; R_4 = H)

Figure 7. The general formulas and names of the main benzoic acids.

In the group of benzoic acids the ones that stand out are protocatechuic acids, vanillic acids, syringic acid, gentisic acid, salicylic acid, p-hydroxybenzoic acid and gallic acid [3].

Among the cinnamic acids, p-coumaric, ferulic, caffeic and sinapic acids are the most common in nature [24].

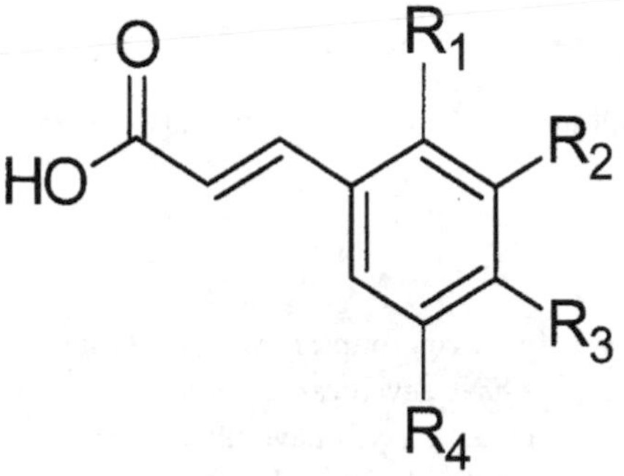

Ceramic acid $(R_1 = R_2 = R_3 = R_4 = H)$
o-Coumaric acid $(R_1 = OH; R_2, R_3, R_4 = H)$
m-Coumaric acid $(R_2 = OH; R_1, R_3, R_4 = H)$
p-Coumaric acid $(R_3 = OH; R_1, R_2, R_4 = H)$
Caffeic acid $(R_2 = R_3 = OH; R_1, R_4 = H)$
Ferulic acid $(R_2 = OCH_3; R_3 = OH; R_1, R_4 = H)$
Sinapic acid $(R_2 = R_4 = OCH_3; R_3 = OH; R_1 = H)$

Figure 8. The general formulas and names of the main cinnamic acids.

Cinnamic acids rarely found free in plants. They are generally in the form of esters, along with a cyclic alcohol-acid, such as quinic acid to form the isochlorogenic acid, neochlorogenic acid, cripto chlorogenic acid and chlorogenic acid, an caffeoyl ester, which is the most important combination [10].

Figure 9 shows the chemical structure of chlorogenic acid.

Phenolic acids may be about one-third of the phenolic compounds in the human's diet [24]. It is known that these substances and their esters have a high antioxidant activity, especially hydroxybenzoic acid, hydroxycinnamic acid, caffeic acid and chlorogenic acid, and although other characteristics also contribute to the antioxidant activity of phenolic acids and their esters, this activity is usually determined by the number of hydroxyl groups found in the molecule thereof. In general, the hydroxylated cinnamic acids are more effective than their benzoic acids counterparts [16].

Despite the antioxidant activity of phenolic compounds and their possible benefits to human health, until the beginning of the last decade, most studies on these substances occurred in relation to their deleterious effects. Tannins, one of the major components of this group, due to the large number of hydroxyl groups contained therein, among other functional groups (1 to 2 per 100 D), are capable of forming strong complexes with proteins, starch and other molecules, particularly digestive enzymes, reducing the digestibility of the feed. Likewise, by joining with their hydroxyl and carbonyl groups, tannins have the ability to chelate divalent cations, especially Fe and Zn, reducing the bioavailability of these minerals [10].

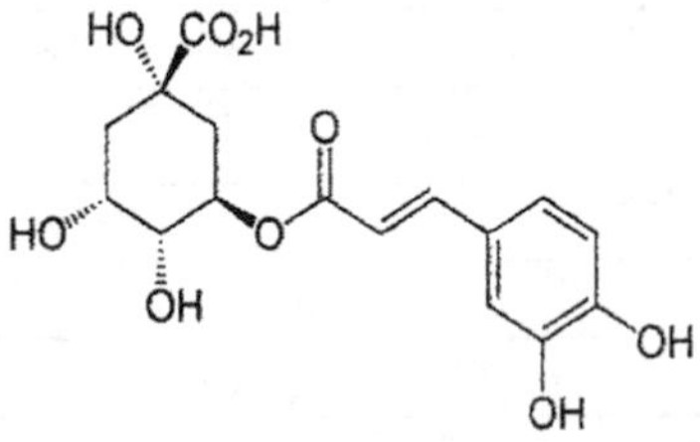

Figure 9. Chemical structure of chlorogenic acid.

Although phenolic compounds are traditionally considered antinutrients, and until the moment as non-nutrients because deficiency states are unknown for them, in recent years they have been seen as a group of micro-nutrients in the vegetable kingdom, which are important part of human and animal diet. The condensed and hydrolysable tannins (ellagitannins) of high molecular weight, since they are not absorbed by the mucosa, they have been regarded as insoluble antioxidants that may have high antioxidant activity in the gastrointestinal tract, protecting proteins, lipids and carbohydrates from oxidative damage during digestion [25].

Researches have also suggested that regular consumption of phenolic compounds directly from plant foods may be more effective in combating oxidative damage in our body than in the form of dietary supplement [26]. This can be explained by the possible synergistic interactions among food phenolic compounds, increasing the antioxidant capacity of these substances..

This way, the content of phenolic compounds and the antioxidant power of a wide variety of plant foods have been investigated.

3. Sources and their antioxidant power

Table 1 shows the mean content of total phenolic compounds (mg/ 100 g of sample) of some plant foods.

Source	Total phenolics (mg%)	Reference
Cereals and legumes		
Cowpea (*V. unguicuata*), brown	100	27
Soyabean	414	28
Oat	352	29
Wheat flour	184	30
Vegetables		
Black carrot	68	31
Broccoli	88	31
Brussels sprouts	69	31
Cabbage, white	76	32
Cabbage, red	186	32
Endive	92	32
Kale	136	33
Lettuce	107	32
Potato	150	31
Spinach	112	32
Tomato	68	32
Yam	92	31
Herbs and spices		
Basil	4425	34
Chilli, green	107	32
Chilli, red	277	32
Coriander	374	31
Garlic	145	31
Ginger	221	31
Leek	85	32
Mint	400	31
Onion, white	269	35
Onion, yellow	164	35
Onion, red	428	35
Pepper, black	1600	36
Pepper, white	800	36

Source	Total phenolics (mg%)	Reference
Shallot	1718	35
Sweet onion	142	35
Thyme	1646	37
Turmeric	176	31
Fruits		
Apple, green	118	38
Apple, red	125	38
Apple, yellow	100	38
Blueberry	362	39
Cherry, sour	156	40
Cherry, sweet	79	38
Grape, black	213	38
Grape, white	184	38
Grapefruit	893	41
Guava, pink flesh	247	42
Guava, white flesh	145	42
Kiwi	791	43
Lemon	843	41
Lime	751	41
Litchi	60	44
Nectarine, white flesh	38	45
Nectarine, yellow flesh	25	45
Orange, sweet	1343	41
Peach, white flesh	53	45
Peach, yellow flesh	35	45
Pear	125	38
Pineapple	94	44
Plum, black	88	44
Plum, red	73	44
Pomegranate	147	44
Pomelo	57	44
Raspberry, black	670	46

Source	Total phenolics (mg%)	Reference
Raspberry, red	342	46
Raspberry, yellow	426	46
Strawberry	199	47
Others		
Roasted cocoa bean	1305	48
Cocoa liquor	994	48
Alkalised cocoa powder	896	48
Baking chocolate	349	48
Red wine	242	49
Tea, black	62	50
Tea, green	83	50
Coffee	188	51

Table 1. Total phenolic compounds content of some plant foods.

As can be seen in Table 1, phenolic compounds are widely distributed in plant foods.

Cocoa, potato, yam, tomato, kale, Brussels sprouts, broccoli and others dark green leafy and brightly-colored vegetables as well as legumes and cereals, in addition to spices and fruits such as cherries and citrus, are particularly rich in phenolic compounds. Red wine also has a high concentration of phenolic compounds. It is known that the abundant phenolic compounds in red wine are anthocyanin [6, 52]. The green and black teas have been extensively studied, since they may contain up to 30% of their dry weight as phenolic compounds [53]. Coffee is also rich in phenolic compounds, especially chlorogenic acid. It has about 7% of the dry weight of the grains [24] and 15% of the dry instant coffee as phenolic compounds [54].

Although in some studies a few statistically significant correlations were found between the levels of total phenolic compounds and antioxidant power of foods, in others the total phenolics content of samples was highly correlated with the antioxidant capacity. On the other hand, there are still no standard methods and approved for determining the antioxidant power *in vitro*. The several available tests for this purpose involve different mechanisms of antioxidant defense system, from the chelation of metal ions to the measure of preventing oxidative damage to biomolecules, and offer distinct numerical results that are difficult to compare. Because of this, studies have used different methods to evaluate the antioxidant capacity of the studied sample, such as ABTS (2,2-azino-bis-3-ethylbenzothiazoline-6-sulfonic acid radical assay), DPPH (2,2-diphenyl-picrylhydrazyl radical assay), FRAP (Ferric Reducing/Antioxidant Power assay) and ORAC (Oxygen Radical Absorbance Capacity assay), among others tests.

By determining the content of total phenolic compounds and ability to reduce $FeCl_3$ as well as DPPH free radical of some commonly consumed and underutilized tropical legumes [27], it was concluded that one of the commonly consumed cowpea *Vigna unguiculata* (brown) as well as underutilized legumes C. cajan (brown) and *S. sternocarpa* could be considered as functional foods due to their relatively higher antioxidant power, which could be as a result of their relative higher total phenolics content. In a similar way, evaluating the antioxidant capacity of twenty soybean hybrids by DPPH assay and their total phenolics content [28], it was concluded that the two cultivars that showed the highest contents of total phenolics also showed the highest antioxidant powers.

Among cereals, milled oat groat pearlings, trichomes, flour, and bran were evaluated as to their antioxidant capacity against the oxidation of R-phycoerythrin protein in the ORAC assay, as well as against the oxidation of low density lipoproteins (LDL) [55]. In both the methods applied the antioxidant capacity of the fractions of oats was in the following order: pearlings > flour > trichome = bran. It was concluded through this study that a part of oat antioxidants, which is rich in phenolic compounds [29], is probably heat-labile because greater antioxidant power was found among the non-steam-treated pearlings. In another study, ten varieties of soft wheat were compared as to their content of total phenolic compounds and antioxidant capacity [30]. Important DPPH, oxygen, hydroxyl and ABTS radical removal capacity was found in all the studied varieties and the content of total phenolics of the samples showed correlation with their antioxidant power in DPPH, ORAC and ABTS assays.

On the other hand, searching the antioxidant capacity of vegetables in the genus Brassica and the best solvent (ethanol, acetone and methanol) for the extraction of their phenolic compounds [56], the results showed that the solvent used significantly affects the phenolics content and the properties of the studied extract. Methanolic extract showed the largest content of total phenolics of broccoli, Brussels sprouts, and white cabbage. In this study, the antioxidant power of the samples was confirmed by different reactive oxygen species and showed to be concentration-dependent. Kale extracts have also been evaluated as to their content of total phenolic compounds and antioxidant capacity [33]. It can be observed that all studied fractions (free and conjugated forms) were able to remove the DPPH radical and that the content of total phenolic compounds of fractions was highly correlated with their antioxidant power.

Herbs and spices are of particular interest, since they have been proved to have high content of phenolic compounds and high antioxidant capacity. The values of Trolox Equivalent Antioxidant Capacity (TEAC) and content of total phenolics were determined for 23 basil accessions [34]. A positive linear relationship was found between the content of total phenolic compounds and the antioxidant power of samples. This study concluded that basils have valuable antioxidant properties for culinary and possible medical application. The concentration of phenolic compounds in peppercorn (black and white), as well as the ability of hydrolyzed and nonhydrolyzed pepper extracts to remove DPPH, superoxide, and hydroxyl radicals [36] were also investigated. The results obtained showed that hydrolyzed and nonhydrolyzed extracts of black pepper contained significantly more phenolic compounds when compared with those of white pepper. For any of these peppers, the hydrolyzed extract contained significantly more phenolic compounds in comparison with the nonhydro-lyzed extract. A dose-dependent effect was observed for all extracts concerning the power of removing free radical and reactive oxygen species, the black pepper extracts being the most effective. This study concluded that the pepper, especially black, which is an important com-ponent in the diet of many sub-Saharan and Eastern countries due to its nutritional impor-tance, can be considered an antioxidant and radical scavenging. However, evaluating the content of phenolic compounds and antioxidant capacity of 14 herbs and spices [37], al-though a significant correlation has been obtained between the phenolics content and anti-oxidant capacity of samples, it was found that the trend of the antioxidant capacity was different according to the method applied. The leaves of the species Piper showed the high-est antioxidant capacity in both methods studied (Folin-Ciocalteu reagent and FRAP meth-od). Yet, the African mango showed the greatest content of free antioxidant by FRAP method, while by Folin method *Piper umbellatum* excelled followed by thyme. This study concluded that the antioxidant power of plant samples should be interpreted with caution when measured by different methods. In spite of that fact, regardless of the method used, the samples were rich in antioxidants.

In addition to the studies already mentioned, the antioxidant capacity of 36 plant extracts was evaluated by the β-carotene and linoleic acid model system [31] and the content of total phenolic compounds of the extracts was determined. Mint, black carrots, and ginger showed high content of total phenolics. The antioxidant capacity calculated as percentage of oxida-tion inhibition ranged from a maximum of 92% in turmeric extracts to a minimum of 12.8% in long melon. Other foods which have high antioxidant capacity (> 70%) were ginger, mint, black carrots, Brussels sprouts, broccoli, yam, coriander and tomato. The antioxidant power of the samples significantly and positively correlated with their content of total phenolic compounds, allowing the conclusion that the plant foods with high content of phenolic com-pounds can be sources of dietary antioxidants. In another study, 66 types of plant foods were analyzed as to their content of phenolic compounds and their antioxidant capacity in the ORAC assay [32]. The results showed that the antioxidants composition and concentra-tion varied significantly among the different vegetables. The coriander, Chinese kale, water spinach and red chili showed high content of total phenolics and high antioxidant power.

Due to the growing recognition of their nutritional and therapeutic value, many fruits have also been investigated as to their content of phenolic compounds and antioxidant capacity. By evaluating the antioxidant capacity and total phenolics content, in addition to flavanol and monomeric anthocyanins, it was found from the flesh and peel of 11 apple cultivars [57] that the concentrations of the parameters investigated differed significantly among the culti-vars and were higher in the peel in comparison to the flesh. The content of total phenolics and antioxidant capacity were significantly correlated in both flesh and peel. It was conclud-ed that the contribution of phenolics to the antioxidant power in apple peel suggests that peel removal may induce a significant loss of antioxidants. It is also known that one of the most important sources of antioxidants among fruits is small red fruits. By determining the antioxidant capacity of four cultivars of blueberry through three different assays (DPPH, ABTS and FRAP), as well as the content of total phenolic compounds, in addition to flavonoids, anthocyanins and flavan-3-ols [39], it was found that all cultivars contained high content of total phenolics, flavonoids and anthocyanins and lower content of flavan-3-ols. However, significant differences were found in the total phenolics content among the different cultivars and growing seasons. Despite this, the studied cultivars showed high antioxidant power, which was highly correlated with the samples phenolic compounds. Similarly, by checking the content of total phenolics, in addition to flavonoids and anthocyanins, as well as the antioxidant capacity of three cultivars of sour cherries [58], a significant difference was observed in phenolics content among different cultivars and growing seasons. However, the cultivars analyzed showed high antioxidant capacity, which was correlated with the phenolic compounds found in them. In this study significant increases were also found in the content of total phenolic compounds and antioxidant power during the ripening of fruits. Additionally, different solvents were applied for comparing the antioxidant capacity and the yield of total phenolic compounds present in the extracts of sour and sweet cherries [40]. It was found that the solubility of phenolic compounds was more effective in extracts of sweet cherries with use of methanol at 50% and in extracts of sour cherries with the use of acetone at 50%. Extracts from lyophilized sour cherries (methanolic and acetone water-mixtures) presented in average twice as high phenolic compounds than ethanolic extracts. The DPPH antiradical efficiency values were higher in the extracts of sour cherries when compared with those of sweet cherries. It was concluded in this work that the strong antioxidant power of extracts of sour cherries is due to the substantial amount of total phenolic compounds present in them and that the fresh sour cherry can be considered as a good dietary source of phenolic compounds. The total phenolics content, total monomeric anthocyanins and antioxidant capacities of 14 wild red raspberry accessions were also examined [59]. In this study, more two cultivars were included in the investigation to determine the variation between wild and cultivated raspberries. Antioxidant capacity of fruits was evaluated by both FRAP and TEAC assays. Significant variability was found for total phenolics, total monomeric anthocyanins and antioxidant capacity of wild raspberries. Nevertheless, the results indicated that some of the wild accessions of red raspberries have higher antioxidant power and phytonutrients content than existing domesticated cultivars. Finally, two strawberry cultivars were studied as to their content of total phenolic compounds and antioxidant capacity in different ripeness stages [47]. It was concluded that despite the berries in general have better taste and be more appreciated at ripe stage, higher contents of total phenolic compounds and antioxidant power were observed at pink stage for both strawberry cultivars studied. Also with respect to the fruits, a less known snake fruit was compared with better known kiwi fruit regarding to their total phenolics content and four radical scavenging (FRAP, ABTS, DPPH and CUPRAC/Cupric Reducing Antioxidant Capacity) ability [43]. It was observed similarity between snake fruit and kiwi fruit in the contents of phenolic compounds as well as antioxidant power in DPPH assay. By this study, it was able to conclude that the two fruits can be applied as antioxidant supplements to the normal diet. Consumption of a combination of both fruits could be recommended in order to obtain the best results. In another study, 25 cultivars, 5 each of white-flesh nectarines, yellow-flesh nectarines, white-flesh peaches, yellow-flesh peaches, and plums at the ripe stage were studied for their total phenolics content and antioxidant capacity by the DPPH and FRAP assays [45]. In descending order, the cultivars presenting higher contents of total phenolics were: white-flesh peaches, plums, yellow-flesh peaches and yellow-flesh nectarines. There was a strong correlation between total phenolics and antioxidant power of nectarines, peaches, and plums. By continuing to study the plum fruits, 20 genotypes of plums were investigated for their antioxidant capacity and total phenolics content [60]. Among the 20 genotypes, a strong correlation was observed between the total phenolics and antioxidant power of the samples, which was determined upon the FRAP assay. It was concluded that phenolic compounds seem to play a significant role in antioxidant value and health benefits of plums. Additionally, Mirabelle plums were examined for their antioxidant capacity by different assays (DPPH, FRAP, ORAC) and total phenolics content [61]. The antioxidant power of the plum peels, flesh and pits reflected the total phenolics content of the samples with efficacy increasing of the order: peels < flesh < pits across the assays. Peel and flesh of six pear cultivars were also investigated for their antioxidant capacity by DPPH assay and total phenolics content [62]. The results obtained showed that the total phenolics content in the peel can be up to 25 times higher than in the flesh. The peel also showed higher antioxidant power. The pomegranate is another fruit that has been researched. Its peel, mesocarp and juice were evaluated for their antioxidant power by TEAC and FRAP assays as well as total phenolics content [63]. It was found not only high correlation between TEAC and FRAP values, but also with the total phenolics content, which was in the following order: mesocarp > peel > juice. This study demonstrated that selection of raw materials (co-extraction of arils and peel) and pressure, respectively, markedly affected the profile and content of phenolics in the pomegranate juices, underlining the necessity to optimise these parameters for obtaining products with well-defined functional qualities. Studies have also been carried out to quantify the total phenolics content and antioxidant capacity of citrus fruits. Comparing the antioxidant properties of peel (flavedo and albedo) and juice of grapefruit, lemon, lime and sweet orange, four different antioxidant assays (DPPH, Reducing Power, β-carotene–linoleate Model System and Thiobarbituric Acid Reactive Substances/TBARS) were applied to the volatile and polar fractions of peels and to crude and polar fraction of juices [41]. Phenolic compounds were among the two main antioxidant substances found in all extracts. Peels polar fractions showed the highest contents in phenolics, which probably contribute to the highest antioxidant power found in these fractions. However, peels volatile fractions showed the lowest antioxidant power. In another experiment, grapefruit and sour orange were extracted with five different polar solvents. The total phenolics content of the extracts was determined and the dried fractions were screened for their antioxidant capacity by four different assays (DPPH, Phosphomolybdenum method, Nitroblue tetrazolium/NBT Reduction and Reducing Power) [64]. All citrus extracts showed good antioxidant capacity. The best correlation between total phenolics and radical scavenging activity was observed by DPPH method. It was concluded that the data obtained clearly established the antioxidant power of the studied citrus fruit extracts. Studying the extraction efficiency of five different solvents on the total phenolics content and antioxidant capacities of pomelo and navel oranges by five antioxidant assays (DPPH, ORAC, ABTS, Phospomolybdenum method and Reducing Power) [65], it was found that the total phenolics content of extracts varied according to the solvent used. Significant differences were also found in antioxidant capacity values via the same method in different solvents, as well as on the antioxidant capacity of each extract via different methods. Nonetheless, the broad range of activity of the extracts led to the conclusion that multiple mechanisms are responsable for the antioxidant power of the samples and clearly indicated the potential application value of the citrus fruits studied. Finally, the study of the content of phenolic compounds and antioxidant power of tropical fruits such as guava has also been conducted. One white-fleshed and three pink-fleshed of guava were analyzed as to their content of total phenolics, in addition to ascorbic acid and total carotenoids, as well as to their antioxidant capacity [42]. The ABTS, DPPH and FRAP assays were used for determining the antioxidant capacity in methanol and dichlorome-thane extracts of the samples, while the ORAC assay was used only for determining it in methanol extracts. The results obtained showed that white pulp guava had more total phenolics and ascorbic acid than pink pulp guava. On the other hand, carotenoids were absent in the white pulp guava. In all antioxidant assays the methanol extracts showed good correlation with the content of total phenolics and ascorbic acid, as well as between them, but showed negative correlation with total carotenoids.

In addition to the aforementioned fruits, in the search for new foods rich in phenolic compounds and high antioxidant capacity, unconventional tropical fruits have been widely researched. Accordingly, the Antilles cherry, Barbados cherry or acerola (1063 mg/100 g), camu-camu (1176 mg/100 g), puçá-preto (868 mg/100 g), assai or açaí (454 mg/100 g) and jaboticaba (440 mg/100 g) showed to be rich in phenolic compounds. When testing the antioxidant capacity of these fruits fresh and dry matter by DPPH assay, it was found an association between their antioxidant power and total phenolics content [66]. Similarly, banana passion fruit (635-1018mg/100 g), cashew (445 mg/100 g) and guava apple (309 mg/100 g) also showed a high total phenolic content when evaluated by FRAP and ABTS assays. The antioxidant power of these fruits showed a strong correlation with their total phenolics content [67].

Other plant-originated foods studied for their content of phenolic compounds and antioxidant capacities are as follows. The cocoa and chocolate liquor antioxidant capacities as well as monomeric and oligomeric procyanidins were studied [68]. The results obtained showed that the procyanidins content was correlated with the antioxidant capacity, which was determined by the ORAC assay as an indicator for potential biological activity of the samples. However, following the changes in total and individual phenolics content as well as antioxidant capacity during the processing of cocoa beans [48], it can be noted that the loss of phenolic compounds and antioxidant capacity of cocoa vary according to the degree of technological processing. The roasting process and cocoa nib alkalization had the greatest influence on the content of phenolic compounds and antioxidant power. The antioxidant capacity of 107 different Spanish red wines, from different varieties of grapes, aging processes and vintages [69] was also investigated by different methods and the results showed that all samples had an important capacity of removing hydroxyl radical and were able to block the superoxide radical, but with 10 times lower intensity. The wines also showed important protective action on biomarkers of oxidative stress. However, few statistically significant correlations were found between the levels of total phenolics and antioxidant power of the wines and the values of these correlations were very low. In another investigation, the antioxidant capacities of three Argentine red wines were evaluated by TEAC and FRAP assays. The correlation between antioxidant capacity and content of phenolic compounds as well as between antioxidant capacity and phenolic profile of samples [49] was determined. It can be noted that the wines showed significant antioxidant capacity. However, no significant correlation was found between their antioxidant capacity and total phenolics content. Nevertheless, the canonical correlation and multiple regression analysis showed that the antioxidant capacity of the samples was highly correlated with their profile of phenolic compounds. The results obtained in this study showed the importance of analyzing the phenolic profile of the sample rather than total phenolics to help understand the differences in the antioxidant power of wines, which should be extended to other food products. Among the alcoholic beverages, antioxidant power has also been reported for whiskey, sake and sherries. [70]. In addition to alcoholic beverages, the free radical-scavenging activity and total phenolic content of commercial tea [50] were determined, finding that green tea contained higher content of phenolic compounds than black tea. The antioxidant capacity per serving of green tea was also much higher than that of black tea. However, comparing the content of total phenolics, flavonoids and antioxidant capacity of black tea, green tea, red wine and cocoa by ABTS and DPPH assays [71], it was found that cocoa contains much higher levels of total phenolics and flavonoids per serving than black tea, green tea and red wine. In the two methods applied, the antioxidant power of the samples per serving was found in the following descending order: cocoa, red wine, green tea and black tea. The content of total phenolic compounds and DPPH and ABTS radical removal capacity of coffee extracts obtained by continuous (Soxhlet 1 h and 3 h) and discontinuous (solid-liquid extraction and filter coffeemaker) methods, many solvents (water, methanol, ethanol and their mixtures), successive extractions and water with different pHs (4.5, 7.0 and 9.5) were also evaluated [72]. The coffee extracts with the highest antioxidant capacity were obtained after extraction with water neutral (pH 7.0) in the filter coffeemaker (24 g spent coffee per 400 mL water). In addition, the drink degreasing and lyophilization of the extract permitted to obtain coffee extract powder with high antioxidant power, which can be used as an ingredient or additive in the food industry with potential for preservation and functional properties.

It is also know that tamarind, canola, sesame, linseed and sunflower seeds are other possible sources of phenolic compounds [73] and have high antioxidant capacity. The antioxidant capacity of the striped sunflower seed cotyledon extracts, obtained by sequential extraction with different polarities of solvents, was determined by three in vitro methods: FRAP, DPPH and ORAC [74]. In the three methods applied, the aqueous extract showed higher antioxidant capacity than the ethanolic. When compared with the synthetic antioxidant Butylated Hydroxyl Toluene (BHT), the antioxidant power of the aqueous extract varied from 45% to 66%, according to the used method. It was concluded in this study that the high antioxidant power found for the aqueous extract of the studied sunflower seed suggests that the intake of this seed may prevent *in vivo* oxidative reactions responsible for the development of several diseases.

4. Conclusion

Phenolic compounds are widely distributed in plant foods (cereals, vegetables, fruits and others), stressing among them the flavonoids, tannins, chalcones, coumarins and phenolic acids. Although some studies have shown few statistically significant correlations between the levels of total phenolics and antioxidant capacity in foods, in others the content of total phenolic compounds was highly correlated with the antioxidant power of samples. Among the plant foods with a high content of phenolic compounds and antioxidant capacity, we can stand out the dark green leafy and brightly-colored vegetables, in addition to cocoa, soyabean, spices and fruits such as cherries and citrus.

Author details

Maria de Lourdes Reis Giada*

Address all correspondence to: mlgiada@nutricao.ufrj.br

Department of Basic and Experimental Nutrition, Institute of Nutrition, Health Sciences Center, Federal University of Rio de Janeiro, Brazil

References

[1] Vickery, M. L., & Vickery, B. (1981). Secondary plant metabolism. *London: MacMillan.*

[2] Scalbert, A, & Williamson, G. (2000). Dietary intake and bioavailability of polyphenols. *Journal of nutrition*, 130, 2073S-2085S.

[3] Sánchez-Moreno, C. (2002). Compuestos polifenólicos: estructura y classificación: presencia en alimentos y consumo: biodisponibilidad y metabolismo. *Alimentaria*, 329, 19-28.

[4] Hollman, P. C. H. (2001). Evidence for health benefits of plant phenols: local or systemic effects? *Journal of the Science of Food and Agriculture*, 81(9), 842-852.

[5] Stafford, H. A. (1983). Enzymic regulation of procyanidin bisynthesis, lack of a flav-3-en-3-ol intermediate. *Phytochemistry*, 22, 2643-2646.

[6] Morton, L. W., Cacceta, R. A. A., Puddey, I. B., & Croft, K. D. (2000). Chemistry and biological effects of dietary phenolic compounds: relevance to cardiovascular disease. *Clinical and Experimental Pharmacology and Physiology*, 27(3), 152-159.

[7] Harborne, J. B. (1980). Plant phenolics. In: Bell EA, Charlwood BV, Archer B. (ed.) Secondary plant products. *Berlin: Springer-Verlag*, 330-402.

[8]   Tomás-Barberán, F. A., & Espín, J. C. (2001). Phenolic compounds and related en-
      zymes as determinants of quality in fruits and vegetables. *Journal of the Science of Food
      and Agriculture*, 81(9), 853-876.

[9]   Harborne, J. B. (1989). Methods in plant biochemistry. In: Dey PM, Harborne JB. (ed.)
      Plant phenolics. *London: Academic Press*, 1.

[10] Bravo, L. (1998). Polyphenols: chemistry, dietary sources, metabolism and nutritional
      significance. *Nutrition Reviews*, 56(11), 317-333.

[11] Lampe, J. W. (1999). Health effects of vegetables and fruit: assessing mechanisms of
      action in human experimental studies. *The American Journal of Clinical Nutrition*, 70,
      475S-490S.

[12] Percival, M. (1998). Antioxidants. *Clinical Nutrition Insights*, 10, 1-4.

[13] Duthie, G. G, Gardner, P. T, & Kyle, J. A. M. (2003). Plant polyphenols: are they the
      new magic bullet? *Proceedings of the Nutrition Society*, 62(3), 599-603.

[14] Hertog, M. G. L., Hollman, P. C. H., & Venema, D. P. (1992). Optimization of a quan-
      titative HPLC determination of potentially anticarcinogenic flavonoids in vegetables
      and fruits. *Journal of Agricultural and Food Chemistry*, 40(9), 1591-1598.

[15] Jovanovic, S. V., Steenken, S., Tosic, M., Marjanovic, B., & Simic, M. G. (1994). Flavo-
      noids as antioxidants. *Journal of the American Chemical Society*, 116(11), 4846-4851.

[16] Sánchez-Moreno, C. (2002). Compuestos polifenólicos: efectos fisiológicos: actividad
      antioxidante. *Alimentaria*, 329, 29-40.

[17] Rice-Evans, C., Miller, N. J., & Paganga, G. (1996). Structure-antioxidant activity rela-
      tionships of flavonoids and phenolic acids. *Free Radical Biology and Medicine*, 20(7),
      933-956.

[18] Chung, K. T., Wong, T. Y., Wei, C. I., Huang, Y. W., & Lin, Y. (1998). Tannins and
      human health: a review. *Critical Reviews in Food Science and Nutrition*, 38(6), 421-464.

[19] Ragan, M. A., & Glombitza, K. (1986). Phlorotannin: Brown algal polyphenols. *Prog-
      ress in Physiological Research*, 4, 177-241.

[20] Okuda, T., Yoshida, T., & Hatano, T. (1995). Hidrolyzable tannins and related poly-
      phenols. *Fortschritte der Chemie organischer Naturstoffe*, 66, 1-117.

[21] Karakaya, S. (2004). Bioavailability of phenolic compounds. *Critical Reviews in Food
      Science and Nutrition*, 44(6), 453-464.

[22] Matern, V., Lüer, P., & Kreusch, D. (1999). Biosynthesis of coumarins. In: Barton D,
      Nakanishi K, Meth-Cohn O, Sankawa V. (ed.) Comprehensive natural products
      chemistry: polyketides and other secondary metabolites including fatty acids and
      their derivatives. *Amsterdam: Elsevier Science*, 623-637.

[23] Pratt, D. E., & Hudson, B. J. F. (1990). Natural antioxidant no exploited commercially.
      In: Hudson BJF. (ed.) Food antioxidants. *London: Elsevier Applied sciences*, 171-180.

[24] Yang, C. S., Landau, J. M., Huang, M. T., & Newmark, H. L. (2001). Inhibition of car-
      cinogenesis by dietary polyphenolic compounds. *Annual Review of Nutrition*, 21,
      381-406.

[25] Martínez-Valverde, I., Periago, M. J., & Ros, G. (2000). Significado nutricional de los
      compuestos fenólicos de la dieta. *Archivos Latinoamericanos de Nutrición*, 50(1), 5-18.

[26] Martin, K. R., & Appel, C. L. (2010). Polyphenols as dietary supplements: A double-
      edged sword. *Nutritional and Dietary Supplements*, 2, 1-12.

[27] Oboh, G. (2006). Antioxidant properties of some commonly consumed and underutil-
      ized tropical legumes. *European Food Research and Technology*, 224(1), 61-65.

[28] Malencić, D., Popović, M., & Miladinović, J. (2007). Phenolic content and antioxidant
      properties of soybean (Glycine max (L.) Merr.) seeds. *Molecules*, 12(3), 576-581.

[29] Kovácová, M., & Malinová, E. (2007). Ferulic and coumaric acids, total phenolic com-
      pounds and their correlation in selected oat genotypes. *Czech Journal of Food Sciences*,
      25(6), 325-332.

[30] Lv, J., Yu, L., Lu, Y., Nui, Y., Liu, L., Costa, J., & Yu, L. (2012). Phytochemical compo-
      sitions, and antioxidant properties, and antiproliferative activities of wheat flour.
      Food Chemistry, doi: 10.1016/j.foodchem.2012.04.141.

[31] Kaur, C., & Kapoor, H. C. (2002). Anti-oxidant activity and total phenolic content of
      some Asian vegetables. *International Journal of Food Science and Technology*, 37(2),
      153-161.

[32] Isabelle, M., Lee, B. L., Lim, M. T., Koh, W. P., Huang, D., & Ong, C. N. (2010). Anti-
      oxidant activity and profiles of common vegetables in Singapore. *Food Chemistry*,
      120(4), 993-1003.

[33] Ayaz, F. A., Hayirhoglu-Ayaz, S., Alpay-Karaoglu, S., Grúz, J., Valentová, K., Ulri-
      chová, J., & Strnad, M. (2008). Phenolic acid contents of kale (Brassica oleraceae L.
      var. acephala DC.) extracts and their antioxidant and antibacterial activities. *Food
      Chemistry*, 107(1), 19-25.

[34] Javanmardi, J., Stushnoff, C., Locke, E., & Vivanco, J. M. (2003). Antioxidant activity
      and total phenolic content of Iranian Ocimum accessions. *Food Chemistry*, 83(4),
      547-550.

[35] Lu, X., Wang, J., Al-Qadiri, H. M., Ross, C. F., Powers, J. R., Tang, J., & Rasco, BA.
      (2011). Determination of total phenolic content and antioxidant capacity of onion (Al-
      lium cepa) and shallot (Allium oschaninii) using infrared spectroscopy. *Food Chemis-
      try*, 129(2), 637-644.

[36] Agbor, G. A., Vinson, J. A., Oben, J. E., & Ngogang, J. Y. (2006). Comparative analysis
      of the in vitro antioxidant activity of white and black pepper. *Nutrition Research*,
      26(12), 659-663.

[37] Agbor, G. A., Oben, J. E., Ngogang, J. Y., Xinxing, C., & Vinson, J. A. (2005). Antioxidant capacity of some herbs/spices from Cameroon: A comparative study of two methods. *Journal of Agricultural and Food Chemistry*, 53(17), 6819-6824.

[38] Marinova, D., Ribarova, F., & Atanassova, M. (2005). Total phenolics and total flavonoids in Bulgarian fruits and vegetables. *Journal of the University of Chemical Technology and Metallurgy*, 40(3), 255-260.

[39] Dragović-Uzelac, V., Savić, Z., Brala, A., Levaj, B., Kovacević, D. B., & Biško, A. (2010). Evaluation of phenolic content and antioxidant capacity of blueberry cultivars (Vaccinium corymbosum L.) grown in the Northwest Croatia. *Food Technology and Biotechnology*, 48(2), 214-221.

[40] Melicháčová, S., Timoracká, M., Bystrická, J., Vollmannová, A., & Céry, J. (2010). Relation of total antiradical activity and total polyphenol content of sweet cherries (Prunus avium L.) and tart cherries (Prunus cerasus L.). *Acta Agriculturae Slovenica*, 95(1), 21-28.

[41] Guimarães, R., Barros, L., Barreira, J. C. M., Sousa, M. J., Carvalho, A. M., & Ferreira, I. C. F. R. (2010). Targeting excessive free radicals with peels and juices of citrus fruits: grapefruit, lemon, lime and Orange. *Food and Chemical Toxicology*, 48(1), 99-106.

[42] Thaipong, K., Boonprakob, U., Crosby, K., Zevallos, L. C., & Byrne, D. H. (2006). Comparison of ABTS, DPPH, FRAP, and ORAC assays for estimating antioxidant activity from guava fruit extracts. *Journal of Food Composition and Analysis*, 19(6), 669-675.

[43] Gorinstein, S., Haruenkit, R., Poovarodom, S., Park, Y. S., Vearasilp, S., Suhaj, M., Ham, K. S., Heo, B. G., Cho, J. Y., & Jang, H. G. (2009). The comparative characteristics of snake and kiwi fruits. *Food and Chemical Toxicology*, 47(8), 184-1891.

[44] Fu, L., Xu, B. T., Xu, X. R., Gan, R. Y., Zhang, Y., Xia, E. Q., & Li, H. B. (2011). Antioxidant capacities and total phenolic contents of 62 fruits. *Food Chemistry*, 129(2), 345-350.

[45] Gil, M. I., Tomás-Barberán, F. A., Hess-Pierce, B., & Kader, AA. (2002). Antioxidant capacities, phenolic compounds, carotenoids, and vitamin C contents of nectarine, peach, and plum cultivars from California. *Journal of Agricultural and Food Chemistry*, 50(17), 4976-4982.

[46] Gansch, H., Weber, C. A., & Lee, C. Y. (2009). Antioxidant capacity and phenolic phytochemiclas in black raspberries. *New York State Horticultural Society*, 17(1), 20-23.

[47] Pineli, L. L. O., Moretti, C. L., dos Santos, S. M., Campos, A. B., Brasileiro, A., Córdova, A. C., & Chiarello, M. D. (2011). Antioxidants and other chemical and physical characteristics of two strawberry cultivars at different ripeness stages. *Journal of Food Composition and Analysis*, 24(1), 11-16.

[48] Jolić, S. M., Redovniković, I. R., Marković, K., Šipušić, Đ. I., & Delonga, K. (2011). Changes of phenolic compounds and antioxidant capacity in cocoa beans processing. *International Journal of Food Science and Technology*, 46(9), 1793-1800.

[49] Baroni, M. V., Naranjo, R. D. D. P., García-Ferreyra, C., Otaiza, S., & Wunderlin, D. A. (2012). How good antioxidant is the red wine ? Comparison of some in vitro and in vivo methods to assess the antioxidant capacity of Argentinean red wines. *LWT-Food Science and Technology*, 47(1), 1-7.

[50] Lee, KW, Lee, HJ, & Lee, CY. (2002). Antioxidant activity of black te Lee a vs. green tea. *Journal of Nutrition*, 132(4), 785.

[51] Natella, F., Nardini, M., Belelli, F., & Scaccini, C. (2007). Coffee drinking induces incorporation of phenolic acids into LDL and increases the resistance of LDL to ex vivo oxidation in humans. *The American Journal of Clinical Nutrition*, 86(3), 604-609.

[52] Pellegrini, N., Serafini, M., Colombi, B., Del Rio, D., Salvatore, S., Bianchi, M., & Brighenti, F. (2003). Total antioxidant capacity of plant foods, beverages and oils consumed in Italy assessed by three different in vitro assays. *Journal of Nutrition*, 133(9), 2812-2819.

[53] Thiagarajan, G., Chandani, S., Sundari, C. S., Rao, S. H., Kulkarni, A. V., & Balasubramanian, P. (2001). Antioxidant properties of green and black tea, and their potential ability to retard the progression of eye lens cataract. *Experimental Eye Research*, 73(3), 393-401.

[54] King, A., & Young, G. (1999). Characteristics and occurrence of phenolic phytochemicals. *Journal of the American Dietetic Association*, 99(2), 213-218.

[55] Handelman, G. J., Cao, G., Walter, M. F., Nightingale, Z. D., Paul, G. L., Prior, R. L., & Blumberg, J. B. (1999). Antioxidant capacity of oat (Avena sativa L.) extracts. 1. Inhibition of low-density lipoprotein oxidation and oxygen radical absorbance capacity. *Journal of Agricultural and Food Chemistry*, 47(12), 4888-4893.

[56] Jaiswal, A. K., Abu-Ghannam, N., & Gupta, S. (2012). A comparative study on the polyphenolic content, antibacterial activity and antioxidant capacity of different solvent extracts of Brassica oleracea vegetables. *International Journal of Food Science and Technology*, 47(2), 223-231.

[57] Vieira, F. G. K., Borges, G. S. C., Copetti, C., Pietro, P. F., Nunes, E. C., & Fett, R. (2011). Phenolic compounds and antioxidant activity of the apple flesh and peel of eleven cultivars grown in Brazil. *Scientia Horticulturae*, 128(3), 261-266.

[58] Mitić, M. N., Obradović, M. V., Kostić, D. A., Micić, R. J., & Pecev, E. T. (2012). Polyphenol content and antioxidant activity of sour cherries from Serbia. *Chemical Industry and Chemical Engineering*, 18(1), 53-62.

[59] Çekiç, Ç., & Özgen, M. (2010). Comparison of antioxidant capacity and phytochemical properties of wild and cultivated red raspberries (RubusidaeusL.). *Journal of Food Composition and Analysis*, 23(6), 540-544.

[60] Rupasinghe, H. P. V., Jayasankar, S., & Lay, W. (2006). Variation in total phenolics and antioxidant capacity among European plum genotypes. *Scientia Horticulturae*, 108(3), 243-246.

[61] Khallouki, F., Haubner, R., Erben, G., Ulrich, C. M., & Owen, R. W. (2012). Phytochemical composition and antioxidant capacity of various botanical parts of the fruits of Prunus × domestica L. from the Lorraine region of Europe. *Food Chemistry*, 133(3), 697-706.

[62] Sánchez, A. C. G., Gil-Izquierdo, A., & Gil, M. I. (2003). Comparative study of six pear cultivars in terms of their phenolic and vitamin C contents and antioxidant capacity. *Journal of the Science of Food and Agriculture*, 83(10), 995-1003.

[63] Fischer, U. A., Carle, R., & Kammerer, D. R. (2011). Identification and quantification of phenolic compounds from pomegranate (Punica granatum L.) peel, mesocarp, aril and differently produced juices by HPLC-DAD-ESI/MS. *Food Chemistry*, 122(2), 807-821.

[64] Jayaprakasha, G. K., Girennavar, B., & Patil, B. S. (2008). Radical scavenging activities of Rio Red grapefruits and Sour orange fruit extracts in different in vitro model systems. *Bioresource Technology*, 99(10), 4484-4494.

[65] Jayaprakasha, G. K., Girennavar, B., & Patil, B. S. (2008). Antioxidant capacity of pummelo and navel oranges: Extraction efficiency of solvents in sequence. *Lebenson Wiss Technology*, 41(3), 376-384.

[66] Rufino, M. S. M., Alves, R. E., Brito, E. S., & Pérez-Jiménez, J. (2010). Bioactive compounds and antioxidant capacities of 18 non-traditional tropical fruits from Brazil. *Food Chemistry*, 121(4), 996-1002.

[67] Contreras-Calderón, J, Calderón-Jaimes, L, Guerra-Hernández, E, & García-Villanova, B. (2011). Antioxidant capacity, phenolic content and vitamin C in pulp, peel and seed from 24 exotic fruits from Colombia. *Food Research International*, 44(7), 2047-2053.

[68] Adamson, G. E., Lazarus, A. S., Mitchell, A. E., Prior, R. L., Cao, G., Jacobs, P. H., Kremers, B. G., Hammerstone, J. F., Rucker, R. B., Ritter, K. A., & Schmitz, H. H. (1999). HPLC method for the quantification of procyanidins in cocoa and chocolate samples and correlation to total antioxidant capacity. *Journal of Agricultural and Food Chemistry*, 47(10), 4184-4188.

[69] Rivero-Pérez, M. D., Muñiz, P., & González-Sanjosé, M. L. (2007). Antioxidant profile of red wines evaluated by total antioxidante capacity, scavenger activity, and biomarkers of oxidative stress methodologies. *Journal of Agricultural and Food Chemistry*, 55(14), 6476-5483.

[70] Moure, A., Cruz, J. M., Franco, D., Domínguez, J. M., Sineiro, J., Domínguez, H., Núñez, M. J., & Parajó, J. C. (2001). Natural antioxidants from residual sources. *Food Chemistry*, 72(2), 145-171.

[71] Lee, K. W., Kim, Y. J., Lee, H. J., & Lee, C. Y. (2003). Cocoa has more phenolic phyto-chemicals and a higher antioxidant capacity than teas and red wine. *Journal of Agricultural and Food Chemistry*, 51(25), 7292-7295.

[72] Bravo, J., Monente, C., Juániz, I., Peña, M. P., & Cid, C. (2011). Influence of extraction process on antioxidant capacity of spent coffee. *Food Research International*, doi: 10.1016/j.foodres.2011.04.026.

[73] Duthie, G. G., Duthie, S. J., & Kyle, J. A. M. (2000). Plant polyphenols in cancer an heart disease: implications as nutritional antioxidants. *Nutrition Research Reviews*, 13(1), 79-106.

[74] Giada, M. L. R., & Mancini-Filho, J. (2009). Antioxidant capacity of the striped sunflower (Helianthus annuus L.) seed extracts evaluated by three in vitro methods. *International Journal of Food Sciences and Nutrition*, 60(5), 395-401.

Chapter 6

Disease and Therapy: A Role for Oxidants

Eva María Molina Trinidad,
Sandra Luz de Ita Gutiérrez,
Ana María Téllez López and Marisela López Orozco

Additional information is available at the end of the chapter

1. Introduction

Oxidative stress is a large increase reduction potential in cell or a decrease in reducing capacity of the cellular redox couples such as glutation. Effects of oxidative stress depend on the magnitude of these changes, if the cell is able to overcome small perturbations and regain its original state. However, severe oxidative stress can cause cell death and even moderate oxidation can trigger apoptosis, whereas if it is too intense can cause necrosis.

A particularly destructive aspect of oxidative stress is the production of reactive oxygen species, which include free radicals and peroxides. Some of the less reactive species (superoxide) can be converted by a redox reaction with transition metals or other compounds quinines redox cycle, more aggressive radical species which can cause extensive damage cellular. Most of these species derived from oxygen are produced at a low level in normal aerobic metabolism and the damage they cause to cells is constantly repaired. However, under the severe levels of oxidative stress that causes necrotic damage produces ATP depletion prevents cell death by apoptosis control.

The antioxidants are substances that may protect your cells against the effects of free radicals. Free radicals are molecules produced when your body breaks down food, or by environmental exposures like tobacco smoke and radiation. Free radicals can damage cells, and may play a role in heart disease, cancer and other diseases.

Antioxidant substances include beta-carotene, lutein, lycopene, selenium, vitamin A; and vitamin C. Antioxidants are found in many foods. These include fruits and vegetables, nuts, grains, and some meats, poultry and fish.

Free radicals damage may lead to cancer. Antioxidants interact with and stabilize free radicals and may prevent some of the damage free radicals might otherwise cause.

Studies in cancer cells *in vitro* and *in vivo* animal's models suggest that the use of free radicals decreases the growth of malignant cells. However, information from recent clinical trials is less clear. In recent years, large-scale, randomized clinical trials reached inconsistent conclusions.

Clinical trials published in the 1990s reached differing conclusions about the effect of antioxidants on cancer. The studies examined the effect of beta-carotene and other antioxidants on cancer in different patient groups. However, beta-carotene appeared to have different effects depending upon the patient population, therefore it is important to personalize treatment, and we must take into account the variability to treatment and individualize or personalize therapy. Studies made by Blot WJ et al., in 1993 for the treatment of cancer published in Chinese Cancer Prevention Study, investigated the effect of a combination of beta-carotene, vitamin E, and selenium on cancer in healthy Chinese men and women at high risk for gastric cancer. The study showed a combination of beta-carotene, vitamin E, and selenium significantly reduced incidence of both gastric cancer and cancer overall.

A 1994 cancer prevention study entitled the Alpha-Tocopherol (vitamin E)/ Beta-Carotene Cancer Prevention Study (ATBC) demonstrated that lung cancer rates of Finnish male smokers increased significantly with beta-carotene and were not affected by vitamin E. Epidemiologic evidence indicates that diets high in carotenoid-rich fruits and vegetables, as well as high serum levels of vitamin E (alpha-tocopherol) and beta carotene are associated with a reduced risk of lung cancer. Another study made by Omenn GS in 1994, the Beta-Carotene and Retinol (vitamin A). Efficacy Trial (CARET) also demonstrated a possible increase in lung cancer associated with antioxidants.

The 1996 Physicians' Health Study I (PHS) found no change in cancer rates associated with beta-carotene and aspirin taken by U.S. male physicians.

The 1999 Women's Health Study (WHS) made by Lee IM, tested effects of vitamin E and beta-carotene in the prevention of cancer and cardiovascular disease among women age 45 years or older. Among apparently healthy women, there was no benefit or harm from beta-carotene supplementation. Investigation of the effect of vitamin E is ongoing.

Three large-scale clinical trials continue to investigate the effect of antioxidants on cancer. The Women's Health Study (WHS) is currently evaluating the effect of vitamin E in the primary prevention of cancer among U.S. female health professionals age 45 and older.

In 2006, the Selenium and Vitamin E Cancer Prevention Trial (SELECT) is taking place in the United States, Puerto Rico, and Canada. SELECT is trying to find out if taking selenium and/or vitamin E supplements can prevent prostate cancer in men age 50 or older. Also the experimental and epidemiologic investigations suggest that alpha-tocopherol (the most prevalent chemical form of vitamin E found in vegetable oils, seeds, grains, nuts, and other foods) and beta-carotene (a plant pigment and major precursor of vitamin A found in many yellow, orange, and dark-green, leafy vegetables and some fruit) might reduce the risk of cancer, particularly lung cancer. The initial findings of the Alpha-Tocopherol, Beta-Carotene Cancer Prevention Study (ATBC Study) indicated, however, that lung cancer incidence was increased among participants who received beta-carotene as a supplement. Similar results were recently reported by the Beta-Carotene and Retinol Efficacy Trial (CARET), which tested a combination of beta-carotene and vitamin A.

The Physicians' Health Study II (PHS II) is a follow up to the earlier clinical trial by the same name. The study is investigating the effects of vitamin E, C, and multivitamins on prostate cancer and total cancer incidence. In another case the supplementation with alpha-tocopherol or beta-carotene does not prevent lung cancer in older men who smoke. Beta-Carotene supplementation at pharmacologic levels may modestly increase lung cancer incidence in cigarette smokers, and this effect may be associated with heavier smoking and higher alcohol intake.

Antioxidants neutralize free radicals as the natural by-product of normal cell processes. Free radicals are molecules with incomplete electron shells which make them more chemically reactive than those with complete electron shells. Exposure to various environmental factors, including tobacco smoke and radiation, can also lead to free radical formation. In humans, the most common form of free radicals is oxygen. When an oxygen molecule (O_2) becomes electrically charged or "radicalized" it tries to steal electrons from other molecules, causing damage to the DNA and other molecules. Over time, such damage may become irreversible and lead to disease including cancer. Antioxidants are often described as "mopping up" free radicals, meaning they neutralize the electrical charge and prevent the free radical from taking electrons from other molecules.

Because of the importance that involves using antioxidants as an alternative in the treatment and prevention of chronic degenerative diseases is useful to express the potential in the use and development of new drugs that include antioxidants.

Free radicals are highly reactive chemical species that possess an unpaired electron. Due to it is reactivity, the radicals react readily with other molecules. When free radicals come into contact with the molecules of the human body such as proteins, lipids, carbohydrates, DNA nucleic acids, react with them. These reactions cause changes in the normal functions of these primary metabolites, which cause severe damage that can cause diseases such as cancer and degenerative diseases like Parkinson's disease or Alzheimer's disease and atherosclerosis, coronary heart disease and diabetes [1-4].

When any of these afore mentioned diseases, the patient receive the treatment used to treat the particular disease, however, prevention plays a big role. Oxidation in the body tissues caused by free radicals can be prevented with a daily intake of foods that have antioxidants.

The implications of modern life cause changes in eating habits of people, these results in a lack of antioxidants in the body to cope with free radicals that are in contact. The role of antioxidants is to react with free radicals and thus prevent, to react with the primary metabolites, thus acting as natural shields against diseases like cancer [5, 6].

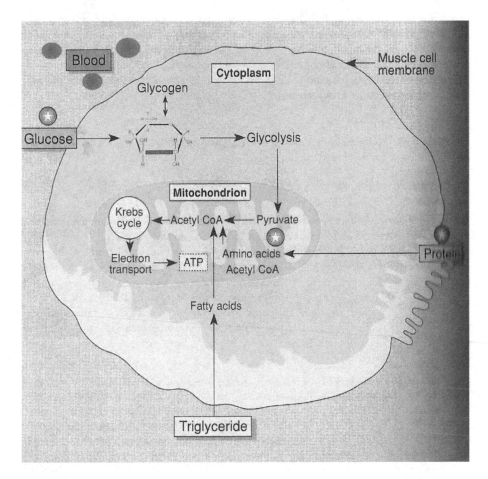

Figure 1. Antioxidants decrement oxidative processes. [Bruce Ames Ph.D., University of California Lecture U.C.T.V. viewed on 08-14-2004].

2. Cancer

2.1. Breast cancer

Currently breast cancer is a disease of high incidence worldwide and causes millions of deaths annually [7]. In the treatment of various cancers have been used drugs that originate from natural products. To get to the application of the drug as a treatment, it requires years of research. The use of treatment leads to the destruction of cancer cells and normal cells in addition, there are numbers side effects resulting from the application of therapies. Prevention of disease is certainly a great alternative to the aggressive use of medications commonly used. The simplest method to prevent cancer and other diseases is undoubtedly add to the diet foods that contain high concentrations of antioxidants, this treatment is easy to perform and causes no adverse side effects. Other organisms containing large amounts of secondary metabolites some of which can act as antioxidants and thereby help prevent cancer and prevent its development (Fruits, vegetables, plants).

Antioxidants can act in two ways:

Blocking cancer, in the initial stage protecting cells against oxidative species and enhancing DNA repair.

Suppressing cancer by inhibiting the progressive stages after formation of pre-neoplastic cells [8].

Studies are underway to help better understand the mechanism of action of antioxidants and test its efficacy against cancer and other diseases. Several studies report that the addition to the diet of foods containing antioxidants may increase the effectiveness of cancer treatment, and help strengthen the body against the side effects associated with treatment [9-11]. The antioxidants found in fruits and vegetables can mention vitamins C and E, carotenoids group and the group of polyphenols. Polyphenols are a group of antioxidant flavonoids to which they belong. There are several types of flavonoids and can be found in foods such as blackberries, blueberries, strawberries, plum, peach, apple, tomato, cherry, broccoli, onion, soya been, legumes like green gram, lupine peas, soy beans, white and horse gram, green leafy spices, citrus fruits, tea, red grapes, chocolate, cocoa and red wine beverages [12].

The following briefly discuss some results of studies using antioxidants from fruits and vegetables for the treatment of breast cancer.

Research conducted in Canada by Hakimuddin and colleagues [13] showed that the polyphenols found in red wine have selective toxicity against MCF-7 cell type of breast cancer; the authors indicate the importance of a diet that incorporates red wine and feeding grapes to serve as a preventive strategy against cancer, which also can be combined with standard therapies.

As mentioned above plums and peaches are fruits that contain phenolic compounds. In a study to test the activity of phenolic species as cancer chemopreventive agents present in extracts of plums and peaches, we found that peaches and plums contain a mixture of phenolic compounds with the ability to inhibit cell lines MCF-7 and MDA -MB435. A very important point to consider is that phenolic acids were isolated chlorogenic and neo-chlorogenic which have great potential for use as chemopreventive agents exerting growth inhibition of the cell line MDA-MB-435 and low toxicity to the normal cell line MCF-10A [14].

Another recent study [15], focused on the action of terpenes located in the skin of the olives suggests that they may serve as natural potential protective against breast cancer. The triterpenes were isolated in significant quantities from the pulp of the olive oil and can act prophylactically and therapeutically.

Currently the investigation for the treatment of breast cancer using apigenin, a flavonoid found in celery. The study conducted at the University of Missouri (United States) [16] was performed in mice that were implanted cell line BT-474, of rapid growth. Mice were also treated with medroxyprogesterone acetate (MPA), which is used in postmenopausal women. Another group of mice was used as a blank. The group of mice treated with MPA was injected apigenin, found that cancerous tumors grew rapidly in mice that were treated with apigenin. Moreover, in mice treated with apigenin was observed a decrease of the tumor when compared with the group of mice used as a blank. Yet unknown mechanism of action of apigenin chemical, however, although the study was conducted in mice, is very promising for future treatment of breast cancer.

2.2. Prostate cancer

Prostate cancer is a very common type of cancer afflicting men; it is now easy detection by prostate specific antigen test (PSA for its initials in English) [17]. Which are still unknown factors that cause this type of cancer, the disease also takes years in some cases to express symptoms, making it necessary for men to undergo regular medical examinations to detect early. One form of treatment of prostate cancer is surgery, whereby the prostate is removed, but this is a procedure which results in urinary incontinence and impotence, which in some cases is permanent.

Prevention through diet prostate cancer has increased because it is recognized as a way to combat this disease [18, 19]. Among the foods that are recommended for the prevention of prostate cancer are generally fruits and vegetables due to its high content of antioxidants. Fruits like pomegranate containing metabolites such as polyphenols and delphinidin urolitina A and B chloride, kaempferol, and punicic acid are considered biologically active against prostate cancer [20, 21].

Other fruit that contains a variety of polyphenolic compounds is strawberry [22] has been found that extracts of strawberry juice cell lines tested against prostate cancer proved effective as antiproliferative agents, is also noteworthy to mention that were tested individually some of the individual components of the extract (cyanidin-3-glucoside, pelargonidin, pelargonidin-3-glucoside, pelargonidin-3-rutinoside, kaempferol, quercetin, kaempferol-3-(6'-coumaroyl) glucoside, 3,4,5 trihydroxyphenyl-acrylic acid-, glucose ester of (E)-p-coumaric acid, and ellagic acid) which also showed efficacy individually [23]. These studies confirm the effectiveness of the cutter to inhibit growth of cancer cells.

The apple is considered the quintessential fruit of health, its daily intake is associated with low risk of chronic diseases and cancer, particularly prostate and colon [24-26]. The block contains a variety of compounds polyphenolic that are responsible for their biological activity among these compounds, studies were performed with quercetin which has proven effective as an inhibitor *in vitro* cell growth of prostate cancer [23, 24]. Another study showed that the antioxidant activity of apples is correlated [27] with the total concentration of phenolic compounds present in it clear that this concentration varies according to growing region, and other growth period factors [28-30]. The tomato is another fruit with high antioxidant capacity and owes its activity to lycopene, a carotenoid, which gives the characteristic red color to the fruit [31, 32]. It has been reported that tomato consumption reduces the occurrence of prostate cancer [33-35].

Another study used extracts of potato species Solanum jamesii to test their cytotoxic activity toward antiproliferatva and prostate cancer cells and colon *in vitro*. The extracts were found to inhibit proliferation of cancer cells PC-3 prostate as well as in colon cancer cells LNCaP. Fractions were also tested extract containing anthocyanin and it showed the same activity as the full extract [36].

2.3. Cervical cancer

It is a type of cancer that has one of the top female deaths worldwide [37]. Its main cause is due to Human Papilloma Virus, which is a group of more than 150 types of viruses and is transmitted by sexual contact [38]. To the treatment of cervical cancer, chemotherapy and radiation therapy is performed. As prevention against this type of cancer was recommended not realize sexual contact with infected persons. Another form of prevention is the application of the vaccine that protects against types of HPV high risk of developing cancer. These vaccines Gardasil ® and Cervarix ® were approved by the Federal Drug Administration (FDA) of EU, but these vaccines are only for women, 9 to 26 years of age who are not infected by the virus.

Another recommendation to prevent this cancer is to stimulate the immune system by eating foods rich in antioxidants, because if the body is weakened, the virus is an opportunity to attack and develop cancer [38]. Have also been performed *in vitro* studies to observe foods as antioxidants influence on the growth of cervical cancer cells [39]. One study was carried out with extracts of different types of berries and tested for anti-proliferative activity on HeLa cells (cervical carcinoma). The results show that extracts from blueberry and pomegranate have little effect inhibiting the growth of HeLa cells. The most effective extracts with increasing concentration were: strawberry extract, arctic bramble, lingonberry and cloudberry. It has also been reported [40] that glycoalkaloids present in commercial potatoes inhibit the growth of different types of cancer cell lines, including HeLa cervical cancer cells.

In therapy of cancer selenium doses is 4000 µg in continuous infusion of 1000 µg/9 days, total: 13 mg [41] (Forceville et al, 2007), i.v. bolus 1000 µg in 30 minutes for continuous infusion 1000 µg/d 14 d, total: 15 mg; i.v. bolus 2000 µg in 2 hours continuous infusion 1600 µg/d, 10 d, total 18 mg [42].

3. Diabetes

Diabetes is a metabolic disorder associated with defects in secretion and insulin action [43]. Type 1 diabetes also known as insulin dependent and type 2 diabetes called non-insulin dependent. Both conditions are associated with the formation of free radicals that cause oxidative stress and disease manifestation. Diabetes is associated with health problems such as neuropathy, retinopathy, erectile dysfunction in men, kidney problems, healing and more

[44, 45]. Because diabetes is a disease of oxidative stress, it is expected that the antioxidants in fruits, vegetables and plants to help combat it.

Several studies report that a proper diet that includes antioxidants is important to reduce the risk of diabetes. We have found that various antioxidants present in some foods and plants as coumarins, some terpenes, flavonoids, lignans, phenylpropanoids, tannins and can help people prevent disease and for helping diabetics [46, 47]. These substances exert their activity by inhibiting the action of R-amylase enzyme. Amylase is an enzyme produced in the pancreas and salivary glands; their function is to help the digestion of carbohydrates [48]. Among the flavonoids that can inhibit R-amylase are the quercetin, myricetin, epigallo-catechin gallate, and cyanidin. Tannins, present in green and black teas, grapes, wine, rasp-berry, and strawberry, also seem to be good R-amylase inhibitors. Among fruits and vegetables reported with inhibitory capacity toward the R-amylase in vitro are the red grapes, strawberry, raspberry and, green pepper, broccoli, ginger, and carrot [49-54].

Thanks to these findings, it has been proposed the use of some natural metabolites present in these fruits for the control of hyperglycemia following ingestion of food. The advantage of these natural metabolites is that its use can avoid the side effects that occur when drugs are used for this purpose [55, 56].

Consumption of foods rich in antioxidants can also prevent the complications of this dis-ease has recently been shown that biotin is a vitamin which is part of the B vitamins, which can be found in foods such as biotin find when we eat certain vegetables: cauli-flower, peanut butter, mushrooms, yeast, potatoes, mushrooms, almonds, walnuts, soy-beans, chickpeas, grapes, strawberries, watermelon, bananas, wheat, flour, pasta, bread, oats, rice, liver, yolk egg, kidney, fish, poultry and offal in general, can help improve metabolism and insulin sensitivity, leading to decreased levels of blood sugar, also sold capsules containing biotin [57, 58].

Resveratrol is a polyphenol present in red wine. According to research Medical Center, University of Texas Southwestern in the U.S. [60], resveratrol administered directly in-to the brain of diabetic mice, can help control type two diabetes by improving blood sugar levels. What makes the resveratrol is to activate a protein called sirtuin which is expressed in parts of the brain that govern the metabolism of glucose. Much remains to be investigated but it is certainly likely that the intake of red wine under medical super-vision can help control diabetes.

Also been studied antioxidants in plants and animals such as the following examples show.

A group of researchers at the University of Jaen in Spain isolated a compound called Cin-namtannin B-1 of the laurel, which has antioxidant properties that can eliminate free radi-cals that cause diseases such as diabetes. The university has signed an agreement with a pharmaceutical for the distribution of this antioxidant [61].

Lipoic acid, also known as alpha lipoic acid or thioctic acid, is produced in small quantities our bodies, it participates in the metabolism significantly. Can also be found in foods like red meat, yeast and some vegetables such as spinach, broccoli.

In this fatty acid properties are attributed as an antioxidant par excellence also can help reuse of other antioxidants like vitamins C and E, glutathione and coenzyme Q10. Among the many properties that are attributed to reduction of varicose veins, skin moisture, enhances energy levels in the body, cancer protection among others.

Also attributed the reduction in blood glucose levels for type 2 diabetes and help combat the discomforts caused by peripheral neuropathy, and therefore coupled with the effects mentioned above, this antioxidant is ideal for diabetics [62-67].

Currently sold in different forms under different names, but the diabetic patient can take doses of lipoic acid consuming identified through the diet. No indication that lipoic acid has contraindications, although high doses can cause episodes of hypoglycemia [68].

4. Arteriosclerosis

Arteriosclerosis is the hardening of the arteries due to fat accumulation; this may lead to a heart attack that can end life [69]. Atherosclerosis is a preventable disease with a balanced diet and exercise. The diet should include variety of fruits and vegetables and be low in fat. Antioxidants play an important role in preventing this disease, it is known that there is a relationship between red wine consumption and the low incidence of cardiovascular disease; this is due to the action of the antioxidants present in grapes. We recommend a daily intake of 375 mL of red wine to increase levels of high density lipoprotein HDL proteins, ie proteins responsible for transporting fat [70, 71]. Studies with another fruits can be determining its effectiveness in the prevention of arteriosclerosis.

Another fruit that has been investigated for its antioxidant and cardiovascular protective effects are blueberries. Studies realized in Arkansas State University, evaluated the effect on two groups of mice for twenty weeks. One group was used as a target, leading a normal diet; the other group was fed a blueberry base [72], found that mice with arterial lesions, a significant percentage decreased injuries, compared with the group of mice that did not eat blueberries. The researchers suggest incorporating blueberries to the diet to improve cardiovascular health and recommended as the ideal fruit for the treatment of hypercholesterolemia.

It is known that fruits such as cranberries have high antioxidant levels and tested their effectiveness in promoting cardiovascular health [73-75]. This study was supplemented to a group of men for two weeks with cranberry juice. Over time he found an increase in plasma antioxidant capacity and a decrease in LDL (low density lipoprotein) in addition to an increase in HDL in obese men. Work is to show whether supplementation based cranberry juice may have the same antioxidant capacity and the same protective benefit as red wine, if so would avoid alcohol.

In another study conducted at the University of Buffalo studied the effect of resveratrol as an antioxidant and its possible use in treating atherosclerosis. In this investigation were not used fruits or vegetables, but was used an extract of the plant. The extract containing resveratrol was administered at doses of 40 mg daily to a group of 10 people, another group of 10 people also served as a target. During the six weeks of the study, blood tests were performed on the results; researchers concluded that Polygonum cuspidatum extract has a therapeutic effect against oxidative stress. These results show that resveratrol, as already mentioned above, are effective to counteract the effect of free radicals, and in the case of arteriosclerosis, can also help prevent it [76].

5. Obesity and metabolic syndrome

The metabolic syndrome has been identified as a target for dietary therapies to reduce risk of cardiovascular disease; however, the role of diet in the etiology of the metabolic syndrome is poorly understood. The metabolic syndrome consists of a constellation of factors that increase the risk of cardiovascular disease and type 2 diabetes. The etiology of this syndrome is largely unknown but presumably represents a complex interaction between genetic, metabolic, and environmental factors including diet [77-79]. The studies endothelial function by assessing the vascular responses to L-arginine, the natural precursor of nitric oxide it's characterized for the low-grade inflammatory state of patients with the metabolic syndrome by measuring circulating levels of high-sensitivity C-reactive protein (hs-CRP) as well as of interleukins 6 (IL-6), 7 (IL-7), and 18 (IL-18). These proinflammatory ILs have been prospectively associated with thrombotic cardiovascular events [80, 81] or have been suggested to be involved in plaque destabilization [82]. The diet designed to increase consumption of foods rich in phytochemicals, antioxidants, α-linolenic acid, and fiber prevent Metabolic Syndrome.

The diet rich in whole grains, fruits, vegetables, legumes, walnuts, and olive oil might be effective in reducing both the prevalence of the metabolic syndrome and its associated cardiovascular risk. One of the mechanisms responsible for the cardioprotective effect of such a diet may be through reduction of the low-grade inflammatory state associated with the metabolic syndrome. Although weight reduction remains a cornerstone of therapy for the metabolic syndrome, from a public health perspective adoption of a diet rich in phytochemicals, antioxidants, α-linolenic acid, and fiber may provide further benefit on cardiovascular risk, especially in patients who do not lose weight.

If antioxidants play a protective role in the pathophysiology of diabetes and cardiovascular disease, understanding the physiological status of antioxidant concentrations among people at high risk for developing these conditions, such as people with the metabolic syndrome, is of interest. However, little is known about this topic. Because the prevalence of obesity, which is associated with decreased concentrations of antioxidants [83], is high among people with the metabolic syndrome, they are probably more likely to have low antioxidant concentrations. Consequently, our purpose was to examine whether concentrations of several antioxidants are lower among those with than those without the metabolic syndrome.

For example a retinol from the liver, the main storage site for retinol is transported to peripheral tissues by retinol binding protein. Retinol may be released as a retinyl ester; however, when the ability of the liver to store retinol is exceeded or when liver function is impaired [84]. Thus, the higher retinyl ester concentrations among those who did not have the metabolic syndrome may indicate that they consumed larger amounts of vitamin A compared with people who have this syndrome. Our findings may have implications for people with the metabolic syndrome, health care professionals who care for them and researchers who study the metabolic syndrome. People with the metabolic syndrome are at increased risk for diabetes and cardiovascular disease, and a role for oxidative stress in the pathophysiology of these conditions has been postulated. Free radical species is one of the principal mechanisms of action of antioxidants, other mechanisms that affect the pathophysiology of diabetes and cardiovascular disease may be operating as well [83]. The effects of vitamins C and E have received a great deal of interest. Through effects on oxidation of LDL cholesterol concentration, leukocyte adhesion, and endothelial function, vitamins C and E may slow atherosclerosis [86, 87].

6. Liver cirrhosis

Currently the evidence supports the role of nutritional deficiency in Alcoholic Liver Disease (ALD) [88–95]. Lieber and colleagues show that progressive ALD proceeds despite adequate nutrition [96, 97]. The latter hypothesis was based primarily on the observation that baboons fed a nutritionally adequate liquid diet containing ethanol at 50% calories developed nearly the whole spectrum of ALD including cirrhosis. Studies demonstrated profound effects on ethanol-induced liver injury by intake of nutrients such as polyunsaturated fat and iron in quantities that were never thought to be important. The concept of 'sensitization' and 'priming' is currently considered fundamental to our pursuit for elucidation of pathogenetic mechanisms of ALD. The sensitization is a conditioning that makes the target cells, hepatocytes, more vulnerable to harmful effects triggered by ethanol and priming as the effect that promotes specific injurious mechanisms. The sensitizing and priming are rendered by the complex interactions of primary mechanistic factors and secondary risk factors. For example, intake of polyunsaturated fat in ethanol-fed rats, but not in pair-fed controls, results in a synergistic priming effect on induction of cytochrome P4502. E1 (CYP2E1) with consequent oxidative injury to the liver [98]. Conversely, saturated fat prevents this priming effect and abrogates depletion of a mitochondrial pool of glutathione (GSH) [99], one of the most crucial sensitization effects of ethanol on hepatocytes [100]. Iron is another example. Whereas a slight increase in hepatic iron content by dietary iron supplementation is harmless in control rats, it exacerbates alcoholic liver injury via accentuation of oxidative stress [101]. Further, increased iron storage in hepatic macrophages is a potential priming mechanism forenhanced expression of tumor necrosis factor a (TNF-a) in experimental ALD [102] Besides nutritional factors, female gender, age, concomitant intake of other drugs that can induce CYP2E1, hepatitis virus infection, and genetic predisposition are all considered risk factors. Even among the primary mechanistic factors that include acetaldehyde, oxidative stress, immune response, hypoxia, and membrane alterations, there are cross-interactive relationships to render sensitization or priming effects. For instance, acetaldehyde, a potent toxic metabolite of ethanol, induces liver injury via its covalent binding to structural or functional proteins of the cells [103] while promoting oxidative stress via consumption of GSH. In turn, deleterious effects of acetaldehyde-protein adduct formation may be accentuated by oxidative stress since malondialdehyde, a lipid peroxidation end product, can increase the binding affinity of acetaldehyde by 13-fold [104]. The resulting novel hybrid adducts are highly immunogenic and may incite immune response mediated liver injury [105, 106]. Although cellular immune response and inflammation are regarded as independent mechanisms of ALD, they can also lead to oxidative stress via the release of reactive oxygen species (ROS) by NADPH oxidase or action of TNF-a at the electron transport chain in target cells. The multifactorial nature and complex interaction among primary mechanistic factors and between primary and secondary factors appear to be the basis for the heterogeneous response that alcoholics exhibit for ALD. Elucidation of the sensitization and priming mechanisms involving cross-interactions of these factors should allow us to gain insight into the most fundamental question, which is why only a small fraction of alcoholics develop advanced ALD. The experimental models to use for control deletion and addition analyses in order to identify what primary and secondary factors are required for the expression of a particular aspect or whole spectrum of experimental ALD. It is need experts in various disciplines need to work together to provide cutting-edge science for elucidating the precise nature and mechanisms that underlie interactions.

Antioxidants represent a potential group of therapeutic agents for ALD. They likely provide beneficial effects on hepatocytes via desensitization against oxidant stress while inhibiting priming mechanisms for expression of proinflammatory and cytotoxic mediators via suppression of NF-kB [107, 108]. Potential approaches may include cell type-specific targeting of antioxidant therapy and development of modalities for more specific and selective regulation of NF-kB-mediated signaling.

The development of cirrhosis is usually associated with oxidative stress and lipid peroxidation (LPO). Studies in models of cirrhosis to use carbon tetrachloride (CCl_4) inhalation in the rat show several similarities with human cirrhosis. The metabolism of CCl_4 into trichloromethyl ($CCl_3\bullet$) and peroxy trichloromethyl ($\bullet OOCCl_3$) free radicals has been reported to cause hepatotoxic effects, like fibrosis, steatosis, necrosis, and hepatocarcinoma [109, 111].

Some compounds that have been studied as possible protectors against liver cirrhosis are known for their anti-inflammatory and antioxidant properties. Plants contain numerous polyphenols, which have been shown to reduce inflammation and thereby to increase resistance to disease [112]. Quercetin (Q), a polyphenolic flavonoid compound present in large amounts in vegetables, fruits, and tea, exhibits its therapeutic potential against many diseases, including hepatoprotection and the inhibition of liver fibrosis [113–114]. It contains a number of phenolic hydroxyl groups, which have strong antioxidant activity [116, 117]. The average intake varies between countries but is approximately 23 mg/day [118].

By increasing the endogenous antioxidant defenses, flavonoids can modulate the redox state of organisms. The major endogenous antioxidant systems include superoxide dismutase (SOD), catalase (CAT), glutathione reductase (GR), and glutathione peroxidase (GPx), which is essential for the detoxification of lipid peroxides [119-121].

7. Hypertension

Excessive reactive oxygen species (ROS) have emerged as a central common pathway by which disparate influences may induce and exacerbate hypertension. Potential sources of excessive ROS in hypertension include nicotinamide adenine dinucleotide phosphate (NADPH) oxidase, mitochondria, xanthine oxidase, endothelium-derived NO synthase, cyclooxygenase 1 and 2, cytochrome P450 epoxygenase, and transition metals. While a significant body of epidemiological and clinical data suggests that antioxidant-rich diets reduce blood pressure and cardiovascular risk, randomized trials and population studies using natural antioxidants have yielded disappointing results. The reasons behind this lack of efficacy are not completely clear, but likely include a combination of [122] ineffective dosing regimens, [123] the potential pro-oxidant capacity of some of these agents, [124] selection of subjects less likely to benefit from antioxidant therapy (too healthy or too sick), and inefficiency of nonspecific quenching of prevalent ROS versus prevention of excessive ROS production. Antioxidants as vitamins A, C and E, L-arginine, flavonoids, and mitochondria-targeted agents (Coenzyme Q10, acetyl-L-carnitine, and alpha-lipoic acid) can be use to treatment hypertension. Currently exist incomplete knowledge of the mechanisms of action of these agents, lack of target specificity, and potential interindividual differences in therapeutic efficacy preclude us from recommending any specific natural antioxidant for antihypertensive therapy at this time.

Reactive oxygen species (ROS) are generated by multiple cellular sources, including NADPH oxidase, mitochondria, xanthine oxidase, uncoupled endothelium-derived NO synthase, cycloxygenase, and lipoxygenase. The dominant initial ROS species produced by these sources is superoxide (O_2^-). Superoxide is short-lived molecule that can subsequently undergo enzymatic dismutation to hydrogen peroxide. Superoxide can oxidize proteins and lipids, or react with endothelium-derived nitric oxide (NO) to create the reactive nitrogen species peroxynitrite. Peroxynitrite and other reactive nitrogen species can subsequently oxidize proteins, lipids, and critical enzymatic cofactors that may further increase oxidative stress [125]. Hydrogen peroxide produced by enzymatic dismutation of O_2^- can be further convert to highly reactive hydroxyl radical (via Fenton chemistry) that can cause DNA damage. The balance between superoxide production and consumption likely keeps the concentration of O_2^- in the picomolar range and hydrogen peroxide in the nanomolar range [126]. These homeostatic levels of reactive oxygen species appear to be important in normal cellular signaling [127-132] and normal reactions to stressors [133, 134].

Randomized trials employing non-pharmacological dietary interventions emphasizing fruits, vegetables, whole grains, and nuts have shown impressive blood pressure lowering results in both hypertensive and normotensive subjects [135, 136]. Similar interventions demonstrated to reduce cardiovascular morbidity and mortality continue to maintain interest in the potential of isolating specific compounds enriched in these diets that may be responsible for the overall dietary benefits [137].

The dietary components in these studies are high in compounds known to have antioxidant properties leading many to ascribe the benefits of these diets to their increased content of natural antioxidants. However, prior randomized trials and population studies in healthy populations and patients at high risk for cardiovascular events that have employed combinations of some of these natural antioxidants as dietary supplements have, for the most part, shown disappointing results [138-145]. The reasons behind these disappointing results are not completely clear, but likely include a combination of 1) ineffective dosing and dosing regimens 2) the potential pro-oxidant capacity and other potentially deleterious effects of these some of these compounds under certain conditions [146-148], 3) selection of subjects less likely to benefit from antioxidant therapy (too healthy or too sick). Populations at intermediate cardiovascular risk may be better suitable to see effects of antioxidants in shorter term studies [149], 4) inefficiency of non-specific quenching of prevalent ROS versus prevention of excessive ROS production [150, 151].

When considering antioxidant therapy for hypertension, lessons from prior disappointing attempts to reduce blood pressure and cardiovascular risk with antioxidant therapy should be considered. The profile of an ideal agent is outlined in The importance of patient selection is being increasingly recognized in light of emerging data suggesting that antioxidant supplementation in healthy subjects may blunt the protective benefits of aerobic exercise training, suggesting ROS generation can be beneficial under certain circumstances.

Antioxidants neutralize the oxidative processes and modify levels in plasma	
↑ Lipid peroxidation	↑ MDA (TBAR), F2-isoprostane
↑ NO synthesis	↑ Nitrite, nitrate, nitrotyrosine
↓ Circulating antioxidants	↓ Uric acid, protein SH groups, Bilirubin (unconjugated)
	↓ Ascorbic acid, α-tocopherol, β-carotene, lycopene
	↓ Antioxidant enzymes (GSHPx)
	↓ Selenium, zinc
	↓ GSH
Xanthine oxidase activation	↑ Plasma xanthine oxidase

Table 1. Antioxidants neutralize the oxidative processes and modify levels in plasma. [150]

7.1. Antioxidant vitamins

7.1.1. Vitamin A precursors and derivatives

Vitamin A precursors and derivatives are retinoids that consist of a beta-ionone ring attached to an isoprenoid carbon chain. Foods high in vitamin A include liver, sweet potato, carrot, pumpkin, and broccoli leaf. Initial interest in vitamin A-related compounds focused primarily on beta-carotene, given initial promising epidemiological data with respect to its cardioprotective effects and some correlation with higher plasma levels to lower blood pressure in men. However, concerns about beta-carotene's pro-oxidative potential came to light with a report suggesting adverse mitochondrial effects of beta-carotene cleavage products. Further, adverse mortality data with respect to beta-carotene has limited interest in this compound as an effective antihypertensive agent.

Recently, interest in vitamin A derivatives has turned to lycopene, itself a potent antioxidant [152], found concentrated in tomatoes. One small study has shown a reduction in blood pressure with a tomato-extract based intervention (containing a combination of potential anti-oxidant compounds including lycopene) in patients with stage I hypertension, [153] although second study showed no effect in pre-hypertensive patients [154].

7.1.2. Ascorbic acid (Vitamin C)

L-ascorbic acid is a six-carbon lactone and, for humans, is an essential nutrient. In Western diets, commonly consumed foods that contain high levels of ascorbic acid include broccoli, lemons, limes, oranges, and strawberries. Toxicity potential of this compound is low, although an increased risk of oxalate renal calculi may exist at higher doses (exceeding 2 grams/day).

The initial purported mechanisms for the potential benefits of ascorbate supplementation were centered on quenching of single-electron free radicals. Subsequent research has demonstrated that the plasma concentrations of ascorbate required for this mechanism to be physiologically relevant are not attainable by oral supplementation [155]. However, vitamin C can concentrate in local tissues to levels an order of magnitude higher than that of plasma. At this ascorbate may to effectively compete for superoxide and reduce thiols [156]. Recent data also suggest potential suppressive effects of ascorbate on NADPH oxidase activity [157, 158]. Ascorbate appears to have limited pro-oxidant ability. [159].

Ascorbate's anti-hypertensive efficacy has been evaluated in multiple small studies [160-163] but not all, show modest reductions in blood pressure in both normotensive and hypertensive populations. These data also suggest that supplementation has limited effect on systemic antioxidant markers and little additional blood pressure benefits are seen beyond the 500 mg daily dose. Large scale randomized trial data specific to ascorbate supplementation and its effects on hypertension are currently lacking. Data from Heart Protection Study (HPS) suggest no significant mortality from supplementation with 250mg/day of ascorbate supplementation. However, the relatively low dose of ascorbate, use of combination therapy, and high-risk patient population studied in HPS leave unanswered the key questions of appropriate dosing and target. In the inflammatory processes follow next scheme in the therapy antioxidant [164].

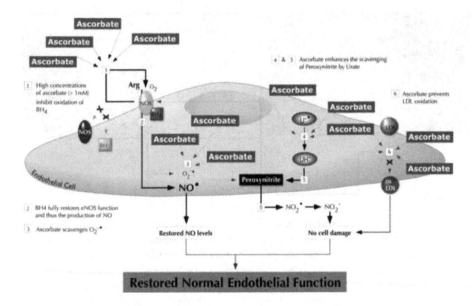

Figure 2. Restored normal endothelial function.

7.1.3. α-Tocopherol (Vitamin E)

Vitamin E is a generic term for a group of compounds classified as tocopherols and tocotrienols [165]. While there are four isomers in each class of Vitamin E compounds, the overwhelming majority of the active form is α-tocopherol. [166, 167]. Dietary sources high in vitamin E include avocados, asparagus, vegetable oils, nuts, and leafy green vegetables.

Vitamin E is a potent antioxidant that inhibits LDL and membrane phospholipid oxidation. Interestingly, inflammatory cells and neurons have binding proteins for α-tocopherol, the actions of which may include inhibition of NADPH oxidase, lipoxygenase, and cyclo-oxygenase, actions which may lower oxidative stress [168]. However, studies demonstrating vitamin E's pro-oxidant capacity under certain cellular conditions suggest that local condition may influence the vitamin E's redox activity [169]. Initial excitement for vitamin E supplementation was based on the reduction of cardiovascular events seen in the CHAOS study. However, follow-up studies have been largely disappointing [170-171]. While one small study that used vitamin E in combination with zinc, vitamin C, and beta-carotene showed a modest, significant reduction in blood pressure over 8 weeks of therapy, other small studies, show either no effect from vitamin E supplementation. Further, the more definitive HOPE trial, failed to show blood pressure or mortality benefit for patients at high risk for cardio-vascular disease [172]. Vitamin E inhibits free radicals reactions.

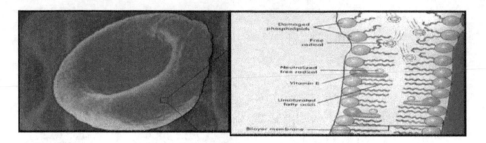

Figure 3. Antioxidants function in the organism.

7.1.4. L-Arginine

L-arginine is an amino acid and the main substrate for the production of NO from eNOS in a reaction that is dependent on tetrahydrobiopterin [173]. Potential dietary sources include milk products, beef, wheat germ, nuts, and soybeans. Reduced levels of tetrahydrobiopterin leads to uncoupling of reduced NADPH oxidation and NO synthesis, with oxygen as termi-nal electron acceptor instead of L-arginine, resulting in the generation of superoxide by eNOS [174-176]. Low cellular levels of L-arginine have been demonstrated in human hyper-tension. While L-arginine deficiency itself does not appear to lead to uncoupling of eNOS, [177] low levels of L-arginine may lead to reduced levels of bioavailable NO which could contribute to hypertension. Thus, L-arginine supplementation could theoretically reduce blood pressure by allowing for restoration of normal NO bioavailability, perhaps overcom-ing overall L-arginine deficiency as well as more successfully competing fo the eNOS active site with circulating asymmetric dimet hylarginine, a circulating competitor of L-arginine that may be increased in the setting of hypertension.

This concept is supported by studies demonstrating the anti-hypertensive effect of L-argi-nine supplementation in salt-sensitive rats, healthy human subjects, hypertensive diabetics, patients with chronic kidney disease, and diabetic patients in combination with N-acetylcys-teine, a precursor of glutathione [178] L-arginine's anti-hypertensive response may be medi-ated in part by its suppressive effects on angiotensin II and endothelin-1, and its potentiating effects on insulin.

However, recent concerns about potential deleterious increases in homocysteine in the set-ting of L-arginine supplementation have been raised. The majority of L-arginine is process-ed into creatine, which leads increased homocysteine levels. Homocysteine can increase oxidative stress. A recent study confirms that this mechanism is relevant to L-arginine me-tabolism in humans [179] suggesting a potential mechanism for neutralizing the eNOS-relat-ed anti-oxidant effects of L-arginine.

7.1.5. Flavonoids

Flavonoids are polyphenolic compounds commonly found in concentrated amounts in multiple fruits, vegetables, and beverages, including apples, berries, grapes, onions, pomegranate, red wine, tea, cocoa, and dark chocolate. The exact structure and composition of the flavonoid compounds varies between food sources, and flavonoid content can be altered based on the manner of food preparation [180]. Interest in flavonoids as antioxidants therapy for cardiovascular disease originates from epidemiological data suggesting improved cardiovascular outcomes in individuals with high intake of food and beverages with high flavonoid content as well as cellular work suggesting a strong anti-oxidant effect of these compounds [181]. However, the limited oral bioavailability of flavonoids suggests cells signaling mechanism, rather than free radical quenching activity, is more likely to be root of sustained cardiovascular benefits from flavonoids [182, 183]. This concept is consistent with studies demondtrating that flavonoids can inhibit NADPH oxidase through ACE inhibition, increase eNOS-specific NO production through the estrogen receptor, and alter COX-2 expression [184]. Studies investigating the anti-hypertensive effects of flavonoids are inconclusive. While multiple small studies of short duration of dark chocolate therapy have demonstrated blood pressure lowering effects in hypertensives [185], studies in normotensive and pre-hypertensive individuals have demonstrated no benefit [186], further tea intake may, at least temporarily, increase blood pressure certain populations [187, 188]. The specific flavonoids and combination of flavonoids that exert the largest beneficial effects remain unknown. The follow table indicates a function of antioxidants in therapy.

Selenium	Septic ICU patients; major burns in combination with Cu and Zn; trauma patients	Ceiling "/>750 µg/day?
Zinc	Pneumonia in children: clinical course significantly shortened	Immune depression if doses"/>50 mg7day are provided
Cu-Se-Zn	Burns: trials showing reduction of infectious complication (pneumonia) and improved wound healing	Doses were calculated to compensate for the exudative losses
Vitamin E (α-tocoferol)	SIRS enteral supplementation	Convincing animal data
Vitamin C (ascorbic acid)	Burns, megadose during the first 24 h after injury; trauma, combined with vitamin E	Possible an endothelial mechanism (189)

Table 2. Antioxidants more indicated in treatments degeneratives chronics.

8. Conclusions

The antioxidants present in food playing an important role in preventing chronic diseases. A balanced diet can prevent diseases associated with oxidative stress and help keep the body in top condition.

Author details

Eva María Molina Trinidad, Sandra Luz de Ita Gutiérrez, Ana María Téllez López and Marisela López Orozco

Universidad Autónoma del Estado de Hidalgo UAEH, Instituto de Ciencias de la Salud IC-Sa, Área Académica de Farmacia. ExHacienda la Concepción, Tilcuautla, Hidalgo, México

References

[1] Huang, X. (2003). Iron overload and its association with cancer risk in humans: evidence for iron as a carcinogenic metal. Mutat.Res. , 533, 153-171.

[2] Markesbery WR, Lovell MA(2006). DNA oxidation in Alzheimer's disease. Antioxid Redox Signal. , 8, 2039-2045.

[3] Halliwell, B. (2001). Role of free radicals in the neurodegenerative diseases: therapeutic implications for antioxidant treatment. Drugs Aging. , 18(9), 685-716.

[4] Vokurkova, M., Xu, S., & Touyz, R. M. (2007). Reactive oxygen species, cell growth, cell cycle progression and vascular remodeling in hypertension. Future Cardiol. Jan; , 3(1), 53-63.

[5] Herrera, E., Jimenez, R., Aruoma, O. I., Hercberg, S., Sanchez-Garcia, I., & Fraga, C. (2009). Aspects of antioxidant foods and supplements in health and disease. Nutr. Rev. 67 (Suppl. 1), SS144., 140.

[6] Dai, J., Jones, D. P., Goldberg, J., Ziegler, T. R., Bostick, R. M., Wilson, P. W., Manatunga, A. K., Shallenberger, L., Jones, L., & Vaccarino, V. (2008). Association between adherence to the Mediterranean diet and oxidative stress. Am. J. Clin. Nutr. , 88, 1364-1370.

[7] Organización Mundial de la Salud(2008). La lucha contra el cáncer tiene que ser una prioridad del desarrollo. Available: http://www.who.int/mediacentre/news/statements/2008/s09/es/index.html.Accessed 2010 December 23.

[8] American Cancer Society. American Cancer Society Cancer Facts and Figures ((2008). Available: http://www.cancer.org/downloads/STT/2008CAFFfinalsecured.pdf Accessed 2009 March 13.

[9] Valko, M., Leibfritz, D., Moncol, J., Cronin, M. T. D., Mazur, M., & Telser, J. (2007). Free radicals and antioxidants in normal physiological functions and human disease. Int. J. Biochem. Cell Biol. j. , 39, 44-84.

[10] Hercberg, S., Galan, P., Preziosi, P., Alfarez, M., & Vazquez, C. (1998). The potential role of antioxidant vitamins in preventing cardiovascular diseases and cancers. Nutrition j. , 14, 513-520.

[11] Borek, C. (2004). Dietary Antioxidants and Human Cancer. Integr. Cancer Ther. , 3, 333-341.

[12] Andreescu, S., et al. (2011). In Oxidative Stress: Diagnostics, Prevention, and Therapy. ACS Symposium Series. American Chemical Society: Washington, DC.

[13] Hakimuddin, F., Paliyath, G., & Meckling, K. (2006). Treatment of Mcf-7 Breast Cancer Cells with a Red Grape Wine Polyphenol Fraction Results in Disruption of Calcium Homeostasis and Cell Cycle Arrest Causing Selective cytotoxicity J. Agric. Food chem. j. 54: (20) 7912-7923.

[14] Noratto, G., Porter, W., Byrne, D., & Cisneros-Zevallos, L. (2009). Identifying peach and plum polyphenols with chemopreventive potential against estrogen-independent breast cancer cells J. Agric. Food chem. j. , 57, 5219-5226.

[15] Allouche, Y., Warleta, F., Campos, M., Sánchez-Quesada, C., Uceda, M., Beltrán, G., & Gaforio, J. J. (2011). Antioxidant, antiproliferative, and pro-apoptotic capacities of pentacyclic triterpenes found in the skin of olives on mcf-7 human breast cancer cells and their effects on DNA damage. J. Agric. Food chem. j. , 59, 121-130.

[16] [16]Available:http://support.dalton.missouri.edu/index.php/daltonnews/ Breast_Cancer_Effectively_Treated_with_Chemical_Found_in_Celery_Parsley_by/. Accessed January 2012.

[17] Heidenreich, A., Aus, G., Bolla, M., Joniau, S., Matveev, V. B., Schmid, H. P., & Zattoni, F. (2008). EAU guidelines on prostate cancer. Eur. Urol. j. 53: (1) 68-80.

[18] Moorthy, H. K., & Venugopal, P. (2008). Strategies for prostate cancer prevention: Review of the literature. Indian J. Urol. j. 24: (3) 295-302.

[19] Singh, R. P., & Agarwal, R. (2006). Mechanisms of action of novel agents for prostate cancer chemoprevention. Endocr.-Related Cancer j. 13: (3) , 751 EOF-78 EOF.

[20] Gonzalez-Sarrias, A., Gimenez-Bastida, J. A., Garcia-Conesa, M. T., Gomez-Sanchez, M. B., Garcia-Talavera, N. V., Gil-Izquierdo, A., Sanchez-Alvarez, C., Fontana-Compiano, L. O., Morga-Egea, J. P., Pastor-Quirante, F. A., Martinez-Diaz, F., Tomas-Barberan, F. A., & Espin, J. C. (2010). Occurrence of urolithins, gut microbiota ellagic acid metabolites and proliferation markers expression response in the human prostate gland upon consumption of walnuts and pomegranate juice. Mol. Nutr. Food Res. j. 54: (3) 311-322.

[21] Gasmi, J., & Sanderson, Thomas. (2010). Growth Inhibitory, Antiandrogenic, and Pro-apoptotic Effects of Punicic Acid in LNCaP Human Prostate Cancer Cells. J. Agric. Food Chem. j. 58: (23) , 12149 EOF-12156 EOF.

[22] Seeram, N. P., Lee, R., Scheuller, H. S., & Heber, D. (2006). Identification of phenolics in strawberries by liquid chromatography electrospray ionization mass spectroscopy. Food Chem. j. , 97, 1-11.

[23] Zhang, Y., Seeram, N. P., Lee, R., Feng, L., & Heber, D. (2008). J. Agric. Food Chem. j. , 56, 670-675.

[24] Willett W C(1995). Diet, nutrition, and avoidable cancer. EnViron. Health Perspect. j. , 103, 165-170.

[25] Eberhardt M V, Lee C Y, Liu R H(2000). Antioxidant activity of fresh apples. Naturej. , 405, 903-904.

[26] Le -Marchand, L., Murphy, S. P., Hankin, J. H., Wilkens, L. R., & Kolonel, L. N. (2000). Intake of flavonoids and lung cancer. J. Natl. Cancer Inst. j. , 92, 154-160.

[27] Xing, N., Chen, Y., Mitchell, S. H., & Young, C. Y. F. (2001). Quercetin inhibits the expression and function of the androgen receptor in LNCaP prostate cancer cells. Carcinogenesisj. , 22, 409-414.

[28] Tsao, R., Yang, R., Xie, S., Sockovie, E., & Khanizadeh, S. (2005). J. Agric. Food Chem. j. , 53(12)

[29] Mc Rae, K. B., Lidster, P. D., de Marco, A. C., & Dick, A. (1990). J Comparison of the polyphenol profiles of the apple fruit cultivars by correspondence analysis. J. Sci. Food Agric. j. , 50, 329-342.

[30] Awad, M. A., de Jager, A., & van Westing, L. M. (2000). Flavonoid and chlorogenic acid levels in apple fruit: characterization of variation. Sci. Hortic. j. , 83, 249-263.

[31] Tsao, R., Yang, R., Young, J. C., & Zhu, H. (2003). Polyphenolic profiles in eight apple cultivars using high-performance liquid chromatography (HPLC). J. Agric. Food Chem. j. , 51, 6347-6353.

[32] Britton, G. (1995). Carotenoids 1: Structure and Properties of Carotenoids in Relation to Function. FASEB J. , 9, 1551-1558.

[33] Di Mascio, P., Kaiser, S., & Sies, H. (1989). Lycopene as the most efficient biological carotenoid singlet oxygen quencher. Arch. Biochem. Biophys. j. , 274, 532-538.

[34] Giovannucci, E., Ascherio, A., Rimm, E. B., Stampfer, M. J., Colditz, G. A., & Willett, Q. C. (1995). Intake of carotenoids and retinol in relation to risk of prostate cancer. J. Natl. Cancer Inst. j. , 87, 1767-1776.

[35] Giovannucci, E. ((1999).) Tomatoes, Tomato-based products, lycopene, and cancer: review of the epidemiological literature. J. Natl. Cancer Inst. j. ., 91, 317-331.

[36] Gann, P. H., Giovannucci, J., Willett, E., Sacks, W., Hennekens, F. M., Stampfer, C. H., & , M. J. (1999). Lower prostate cancer risk in men with elevated plasma lycopene levels: results of a prospective analysis. Cancer Res. j. , 59, 1225-1230.

[37] Reddivari, L., Vanamala, J., Chintharlapalli, S., Safe, S. H., Miller, J. C., & Jr , . (2007). Anthocyanin fraction from potato extracts is cytotoxic to prostate cancer cells through activation of caspase-dependent and caspase-independent pathways. Carcinogenesisj. , 28, 2227-2235.

[38] Available:http://www.cancer.gov/espanol/recursos/hojas-informativas/riesgo-causas/ VPH-respuestas. AccessedFebruary (2012).

[39] Available, http://www.who.int/mediacentre/factsheets/fs297/es/index.html., & Accessed, . March (2012).

[40] Mcdougall, G. J., Ross, H. A., Ikeji, M., & Stewart, D. (2008). Berry Extracts Exert Different Antiproliferative Effects against Cervical and Colon Cancer Cells Grown in Vitro. J. Agric. Food Chem. j. , 56, 3016-3023.

[41] Forceville Xavier, Laviolle Bruno, Annane Djillali, Vitoux Dominique, Bleichner Gérard, Korach Jean Michel, Cantais Emmanuel, Georges Hug.(2007). Effects of high doses of selenium, as sodium selenite, in septic shock: a placebo-controlled, randomized, double-blind, phase II study. Critical Care. http://ccforum.com/content/11/4/ R73,viewed on 27-07-2012., 1-10.

[42] Manzanares, W. ., & Hardy, . Selenium supplementation in the critically ill: posology and pharmacokinetics. G.Curr Opin Clin Nutr Metab Care (2009). , 12, 273-80.

[43] Friedman, M., Lee, K. R., Kim, H. J., Lee, I. S., & Kozukue, N. (2005). Anticarcinogenic effects of glycoalkaloids from potatoes against human cervical, liver, lymphoma, and stomach cancer cells. J. Agric. Food Chem. j. , 53, 6162-6169.

[44] World Health Organization.(1999). Definition, Diagnosis and Classification of Diabetes Mellitus and Its Complications. Part I: Diagnosis and Classification of Diabetes Mellitus; Geneva, Switzerland.

[45] Valko, M., Leibfritz, D., Moncol, J., Cronin, M. T. D., Mazur, M., & Telser, J. (2007). Free radicals and antioxidants in normal physiological functions and human disease. Int. J. Biochem. Cell Biol. j. , 39, 44-84.

[46] Rahimi, R., Nikfar, S., Larijani, B., & Abdollahi, M. (2005). A review on the role of antioxidants in the management of diabetes and its complications. Biomed. Pharmacother. j. , 59, 365-373.

[47] Chu, Y. F., Sun, J., Wu, X., & Liu, R. H. (2002). Antioxidant and antiproliferative activities of common vegetables. J. Agric. Food Chem. j. , 50, 6910-6916.

[48] Liu R H(2004). Potential synergy of phytochemicals in cancer prevention: Mechanism of action. J. Nutr. j. 134: , 3479S EOF-3485S EOF.

[49] Available, http://www.nlm.nih.gov/medlineplus/spanish/ency/article/003464.htm., & Accessed, . January (2012).

[50] Mc Dougall, G. J., Shpiro, F., Dobson, P., Smith, P., Blake, A., & Stewart, D. (2005). Different polyphenolic components of soft fruits inhibit R-amylase and R-glucosidase. J. Agric. Food Chem. j. , 53, 2760-2766.

[51] Tadera, K., Minami, Y., Takamatsu, K., & Matsuoka, T. (2006). Inhibition of R-glucosidase and R-amylase by flavonoids. J. Nutr. Sci. Vitaminol. j. , 52, 149-153.

[52] Mullen, W., Mcginn, J., Lean, M. E. J., Maclean, M. R., Gardner, P., Duthie, G. G., Yo-kota, T., & Crozier, A. (2002). Ellagitannins, flavonoids, and other phenolics in red raspberries and their contribution to antioxidant capacity and vasorelaxation proper-ties. J. Agric. Food Chem. j. , 50, 5191-5196.

[53] Pinto, M. S., Kwon, Y. I., Apostolidis, E., Lajolo, F. M., Genovese, M. I., & Shetty, K. (2008). Functionality of bioactive compounds in Brazilian strawberry (Fragaria x ana-nassa Duch.) cultivars: evaluation of hyperglycemia and hypertension potential us-ing in vitro models. J. Agric. Food Chem. j. , 56, 4386-4382.

[54] Matsui, T., Tanaka, T., Tamura, S., Toshima, A., Tamaya, K., Miyata, Y., Tanaka, K., & Matsumoto, K. (2007). R-Glucosidase inhibitory profile of catechins and theafla-vins. J. Agric. Food Chem. j. , 55, 99-105.

[55] Kwon, Y. I., Vattem, D. A., & Shetty, K. (2006). Clonal herbs of Laminaceae species against diabetes and hypertension. Asia Pac. J. Clin. Nutr. j. , 15, 424-432.

[56] Genovese M I, Pinto M S, Gonc-alves A E S S, Lajolo F M(2008). Bioactive compounds and antioxidant capacity of exotic fruits and commercial frozen pulps from Brazil. Food Sci. Technol. Int. j. , 14, 207-214.

[57] de Souza, A. E., Gonc-alves, S., Lajolo, F. M., & Genovese, M. I. (2010). Chemical Composition and Antioxidant/Antidiabetic Potential of Brazilian Native Fruits and Commercial Frozen Pulps J. Agric. Food Chem. j. DOI:10.1021/jf903875u., 58, 4666-4674.

[58] Available, http://www.lenntech.es/vitaminas/biotina.htm., & Accessed, . February (2012).

[59] Available: http://www.nlm.nih.gov/medlineplus/spanish/druginfo/natural/313.html.

[60] Available: http://www.guia-diabetes.com/el-resveratrol-mejora-la-diabetes-con-su-accion-sobre-el cerebro.html. AccessedJanuary (2012).

[61] Available, http://www.cienciadirecta.com/espanol/web/noticias/ujalaurel9063.asp., & Accessed, . April (2012).

[62] Torissen, O., Hardy, R., & Shearer, K. (1989). Pigmentation of salmonoids carotenoid deposition and metabolism. CRC Crit. ReV. Aq. Sci. j. , 1, 209-225.

[63] Naito, Y., Uchiyama, K., Aoi, W., Hasegawa, G., Nakamura, N., Yoshida, N., Maoka, T., Takahashi, J., & Yoshikawa, T. (2004). Prevention of diabetic nephropathy by treatment with astaxanthin in diabetic db/db mice. Biofactors j. , 20, 49-59.

[64] Jacob, S., Hernrisken, E. J., Schiemann, A. L., et al. (1995). Enhancement of glucose disposal in patients with type 2 diabetes by alpha lipoic acid. Arzeneimittel- For-schung Drug Research j. 45: , 872 EOF-4 EOF.

[65] Lester Packer, Carol Colman(1999). The Antioxidant Miracle: Put Lipoic Acid, Pyco-genol, and Vitamins E and C to Work for You J. ohn Wiley & sons: New York 0-47135-311-6

[66] Allan, E., Sosin, Beth. M., Ley-Jacobs, Julian. M., & Whitaker, . (1998). Alpha Lipoic Acid: Nature's Ultimate Antioxidant Kensington Books. New York. 157566366

[67] Burt Berkson ((1998).) Alpha Lipoic Acid Breakthrough: The Superb Antioxidant That May Slow Aging, Repair Liver Damage, and Reduce the Risk of Cancer, Heart Disease, and Diabetes. Three River Press. New York.

[68] lable:http://www.vitabasix.com/fileadmin/content/produktInfoPDFs/esPDF/Produk-tinfo_ALA_ES.pdf. AccessedMay, (2012).

[69] Available, http://www.nlm.nih.gov/medlineplus/spanish/ency/article/000171.htm., & Accesses, . June, (2012).

[70] Tsang, C., Higgins, S., Duthie, G. G., Duthie, S. J., Howie, M., Mullen, W., Lean, M. E., & Crozier, A. (2005). The influence of moderate red wine consumption on antioxidant status and indices of oxidative stress associated with CHD in healthy volunteers. Br. J. Nutr. j. , 93, 233-240.

[71] Zern T L, Fernandez T L(2005). Cardioprotective effects of polyphenols. J. Nutr. j. , 135, 2291-2294.

[72] Milner J A(2002). Foods and health promotion: The case for cranberry. Crit. ReV. Food Sci. Nutr. j. , 42, 265-266.

[73] Neto C C(2007). Cranberry and blueberry: Evidence for protective effects against cancer and vascular disease. Mol. Nutr. Food Res. j. , 51, 652-664.

[74] Ruel, G., Pomerleau, S., Couture, P., Lamarche, B., & Couillard, C. (2005). Changes in plasma antioxidant capacity and oxidized lowdensity lipoprotein levels in men after short-term CJ consumption. Metabolism j. , 54, 856-861.

[75] Ruel, G., Pomerleau, S., Couture, P., Lemieux, S., Lamarche, B., & Couillard, C. (2006). Favorable impact of low-calorie cranberry juice consumption on plasma HDL cholesterol concentrations in men. Br. J. Nutr. j. , 96, 357-364.

[76] Available, http://www.abajarcolesterol.com/resveratrol-previene-la-ateroesclerosis., & Accessed, . june (2012).

[77] Katherine Esposito, Raffaele Marfella, Miryam Ciotola, Carmen Di Palo, Francesco Giugliano, Giovanni Giugliano, Massimo D'Armiento, Francesco D'Andrea, Dario Giugliano(2004). Effect of a Mediterranean-Style Diet on Endothelial Dysfunction and Markers of Vascular Inflammation in the Metabolic Syndrome. JAMA. , 292(12), 1440-1446.

[78] Groop, L. (2000). Genetics of the metabolic syndrome. Br J Nutr. 83(suppl 1):SS48., 39.

[79] Lidfeldt, J., Nyberg, P., Nerbrand, C., et al. (2003). Socio-demographic and psychological factors are associated with features of the metabolic syndrome: the Women's Health in the Lund Area (WHILA) study. Diabetes Obes Metab. , 5, 106-112.

[80] Harris, T. B., Ferrucci, L., Tracy, R. P., et al. (1999). Association of elevated interleukin-6 and C-reactive protein levels with mortality in the elderly. Am J Med. , 106, 506-512.

[81] Blankenberg, S., Tiret, L., Bickel, C., et al. (2002). Interleukin-18 is a strong predictor of cardiovascular death in stable and unstable angina. *Circulation*, 106, 24-30.

[82] Damâs, J. K., Væhre, T., Yndestad, A., et al. (2003). Interleukin-7-mediated inflammation in unstable angina: possible role of chemokines and platelets. *Circulation*, 107, 2670-2676.

[83] Reitman, A., Friedrich, I., Ben-Amotz, A., & Levy, Y. (2002). Low plasma antioxidants and normal plasma B vitamins and homocysteine in patients with severe obesity. Isr Med Assoc J , 4, 590-593.

[84] Ballew, C., Bowman, Russell. R. M., Sowell, A. L., & Gillespie, C. (2001). Serum retinyl esters are not associated with biochemical markers of liver dysfunction in adult participants in the Third National Health and Nutrition Examination Survey (NHANES III), 1988-1994. Am J Clin Nutr , 73, 934-940.

[85] Visioli, F. (2001). Effects of vitamin E on the endothelium. Equivocal? Alphatocopherol and endothelial dysfunction. Cardiovasc Res , 51, 198-201.

[86] Carr, A. C., Zhu, B. Z., & Frei, B. (2000). Potential antiatherogenic mechanisms of ascorbate (vitamin C) and alpha-tocopherol (vitamin E). Circ Res , 87, 349-354.

[87] Cobo Abreu Carlos.(2001). Acido acetilsalicílico y vitamina E en la prevención de las enfermedades cardiovasculares. Rev Mex Cardiol; , 12(3), 128-133.

[88] Barak, A. J., Tuma, D. J., & Beckenhauer, J. L. (1971). Ethanol feeding and choline defiency as influences on hepatic choline uptake. J. Nutr. , 101, 533-538.

[89] French, S. W. (1966). Effect of chronic ethanol ingestion on liver enzyme changes induced by thiamine, riboflavin, pyridoxine or choline deficiency. J. Nutr. , 88, 291-302.

[90] Gomez-Dumm, C. L. A., Porta, E. A., Hartroft, W. S., & Koch, O. R. (1968). A new experimental approach in the study of chronic alcoholism. II. Effects of high alcohol intake in rats fed diets of various adequacies. Lab. Invest. , 18, 365-378.

[91] Hartfoft, W. S., & Porta, E. A. (1968). Alcohol, diet, and experimental hepatic injury. Can. J. Physiol. Pharmacol. Klatskin, G., Krehl, W. A., and Corn, H. (1954) Effect of alcohol on choline requirement. I changes in rat liver after prolonged ingestion of alcohol. J. Exp. Med. 100, 605-614., 46, 463-473.

[92] Mendenhall, C. L., Anderson, S., Weesner, R. E., Goldberg, S. J., & Crolic, K. A. (1984). Protein-calorie malnutrition associated with alcoholic hepatitis. Veterans Administration Cooperative Study Group on Alcoholic Hepatitis. Am. J. Med. , 76, 211-222.

[93] Porta, E. A., Hartroft, W. S., Gomez-Dumm, C. L. A., & Koch, O. R. (1967). Dietary factors in the progression and regression of hepatic alterations associated with experimental chronic alcoholism. Federation Proc. , 62, 1449-1457.

[94] Takeuchi, J., Takada, A., Ebata, K., Sawaw, G., & Okumura, Y. (1968). Effect of a single intoxication dose of alcohol on the livers of rats fed a choline-deficient diet or a commercial ration. Lab. Invest. , 19, 211-217.

[95] Lieber, C. S., Jones, D. P., Nendelson, J., & De Carli, L. M. (1963). Fatty liver, hyperlipemia and hyperuricemia produced by prolonged alcohol consumption, despite adequate dietary intake. Trans. Assoc. Am. Phys. , 76, 289-300.

[96] Lieber, C. S., De Carli, L. M., & Rubin, E. (1975). Sequential production of fatty liver, hepatitis, and cirrhosis in sub-human primates fed ethanol with adequate diets. Proc. Natl. Acad. Sci. USA , 72, 437-441.

[97] Nanji, A. A., Zhao, S., Lamb, R. G., Dannenberg, A. J., Sadrzadeh, S. M. H., & Wasman, D. J. (1994). Changes in cytochromes B1,4A, phospholipase A and C in intragastric feeding rat model for alcoholic liver disease: relationships to dietary fats and pathologic liver injury. Alcohol. Clin. Exp. Res. 18, 902-908, 4502E1.

[98] Colell, A., Kaplowitz, N., Tsukamoto, H., & Fernandez, Checa. J. C. (1997). Effects of dietary medium chain triglycerides (MCT) on ethanol induced mitochondrial GSH depletion in rat liver and pancreas. J. Hepatol. Suppl. 26, 127.

[99] Colell, A., Garcia-Ruiz, C., Miranda, M., Ardite, E., Mari, M., Morales, A., Corrales, F., Kaplowitz, N., & Fernandez-Checa, J. C. (1998). Selective glutathione depletion of mitochondria by ethanol sensitizes hepatocytes to tumor necrosis factor. Gastroenterology , 115, 1541-1551.

[100] Tsukamoto, H., Horne, W., Kamimura, S., Niemela, O., Parkkila, S., Yla-Herttuala, S., & Brittenham, G. M. (1995). Experimental liver cirrhosis induced by alcohol and iron. J. Clin. Invest. , 96, 620-630.

[101] Tsukamoto, H., Lin, M., Ohata, M., Giulivi, C., French, S., & Brittenham, G. (1999). Iron primes hepatic macrophages for NF-kB activation in alcoholic liver injury. Am. J. Physiol. 277, GG1250., 1240.

[102] Tuma, D. J., Newman, M. R., Donohue, T. M., & Sorrell, M. F. (1987). Covalent binding of acetaldehyde to proteins: participation of lysine residues. Alcohol. Clin. Exp. Res. , 579-584.

[103] Tuma, D. J., Thiele, G. M., Xu, D., Klassen, L. W., & Sorrell, M. F. (1996). Acetaldehyde and malondialdehyde administration. Hepatology , 23, 872-880.

[104] Thiele, G. M., Tuma, D. J., Willis, M. S., Miller, J. A., Mc Donald, T. L., Sorrell, M. F., & Klassen, L. W. (1998). Soluble proteins modified with acetaldehyde and malondialdehyde are immunogenic in the absence of adjuvant. Alcohol. Clin. Exp. Res. , 22, 1731-1739.

[105] Xu, D., Thiele, G. M., Beckenhauer, J. L., Klassen, L. W., Sorrell, M. F., & Tuma, D. J. (1998). Detection of circulation antibodies to malondialdehyde-acetaldehyde adducts in ethanol-fed rats. Gastroenterology , 115, 686-692.

[106] Ingelman-Sundberg, M., & Johansson, I. (1984). Mechanisms of hydroxyl radical formation and ethanol oxidation by ethanol-inducible and other forms of rabbit liver microsomal cytochrome J. Biol. Chem. 259, 6447-6458., 450.

[107] Castillo, T., Koop, D. R., Kamimura, S., Triafilopoulos, G., & Tsukamoto, H. (1992). Pole of cytochrome E1 in ethanol-carbon tetrachloride-and iron-dependent microsomal lipid peroxidation. Hepatology 16, 992-996., 450.

[108] Ekstrom, G., & Ingelman-Sundberg, M. (1989). Tat liver microsomal NADPH-supported oxidase activity and lipid peroxidation dependent on ethanol-inducible cytochrome P-4500IIE1). Biochem. Pharmacol. 38, 1313-1319., 450.

[109] Hill, D. B., Devalaraja, R., Joshi-Barve, S., & Mc Clain, C. J. (1999). Antioxidants attenuate nuclear factor-kappa B activation and tumor necrosis factor-alpha production in a alcoholic hepatitis patient monocytes and rat Kupffer cells, in vitr. o. Clin. Biochem , 32, 563-570.

[110] Mc Clain, C. J., Barve, S., Barve, S., Deaciuc, I., & Hill, D. B. (1998). Tumor necrosis factor and alcoholic liver disease.Alcohol. Clin. Exp. Res. 22, 248S-252S.

[111] Fang, H. L., & Lin, W. C. (2008). Lipid peroxidation products do not activate hepatic stellate cells. Toxicology, 253(1-3), 36-45.

[112] Perez, R., & Tamayo, . (1983). Is cirrhosis of the liver experimentally produced by CCl₄ an adequate model of human cirrhosis? Hepatology, , 3(1), 112-120.

[113] Bengmark, S., Mesa, M. D., Gil, A., & Hernández, . (2009). Plant-derived health-the effects of turmeric and curcuminoids. Nutricion Hospitalaria, 24(3), 273-281.

[114] Amália, P. M., Possa, M. N., Augusto, M. C., & Francisca, L. S. (2007). Quercetin prevents oxidative stress in cirrhotic rats,. Digestive Diseases and Sciences, 52(10), 2616-2621.

[115] González-Gallego, J., Sánchez-Campos, S., & Tuñón, M. J. (2007). Anti-inflammatory properties of dietary flavonoids,. Nutricion Hospitalaria, , 22(3), 287-293.

[116] Tieppo, J., Cuevas, M. J., Vercelino, R., Tuñón, M. J., Marroni, N. P., & González-Gallego, J. (2009). Quercetin administration ameliorates pulmonary complications of cirrhosis in rats. Journal of Nutrition, 139(7), 1339-1346.

[117] Martinez-Florez, S., González-Gallego, J., Culebras, J. M., & Tuñón, M. J. (2002). Flavonoids: properties and anti-oxidizing action. Nutrition Hospital, , 17(6), 271-278.

[118] Tokyol, C., Yilmaz, S., Kahraman, A., Çakar, H., & Polat, C. (2006). The effects of desferrioxamine and quercetin on liver injury induced by hepatic ischaemia-reperfusion in rats. Acta Chirurgica Belgica, 106(1), 68-72.

[119] Abilés, J., Moreno-Torres, R., Moratalla, G., et al. (2008). Effects of supply with gluta-mine on antioxidant system and lipid peroxidation in patients with parenteral nutri-tion," Nutricion Hospitalaria, 23(4), 332-339.

[120] Silvia Bona, Lidiane Isabel Filippin, F´abioCangeri Di Naso, Cintia de David,5 Bruna Valiatti,6 Maximiliano Isoppo Schaun, RicardoMachado Xavier, and Norma Possa-Marroni(2012). Effect of Antioxidant Treatment on Fibrogenesis in Rats with Carbon Tetrachloride-Induced Cirrhosis. International Scholarly Research Network ISRN Gastroenterology. , 2012

[121] Kizhakekuttu TJ, Widlansky ME(2010). Natural antioxidants and hypertension: promise and challenges. Cardiovasc Ther. 28(4):e, 20-32.

[122] Laursen, J. B., Somers, M., Kurz, S., et al. (2001). Endothelial regulation of vasomo-tion in apoE-deficient mice : Implications for interactions between peroxynitrite and tetrahydrobiopterin. *Circulation*, 103(9), 1282-1288.

[123] Munzel, T., Daiber, A., Ullrich, V., & Mulsch, A. (2005). Vascular consequences of en-dothelial nitric oxide synthase uncoupling for the activity and expression of the solu-ble guanylyl cyclase and the cGMP-dependent protein kinase. Arterioscler Thromb Vasc Biol, , 25(8), 1551-1557.

[124] Thomas, S. R., Chen, K., Keaney, J. F., & Jr , . (2002). Hydrogen peroxide activates en-dothelial nitric-oxide synthase through coordinated phosphorylation and dephos-phorylation via a phosphoinositide 3-kinase-dependent signaling pathway. J Biol Chem. 22; , 277(8), 6017-6024.

[125] Drummond, G. R., Cai, H., Davis, Ramasamy. S., & Harrison, D. G. (2000). Transcrip-tional and posttranscriptional regulation of endothelial nitric oxide synthase expres-sion by hydrogen peroxide. Circ Res. 18; , 86(3), 347-354.

[126] Stocker, R., Keaney, J. F., & Jr , . (2004). The role of oxidative modifications in atheros-cle. rosis. Physiol Rev; , 84, 1381-1478.

[127] Chen, K., Thomas, S. R., Keaney, J. F., & Jr Beyond, . (2003). LDL oxidation: ROS in vascular signal transduction. Free Radic Biol Med, 15; , 35(2), 117-132.

[128] Moore TJ, Vollmer WM, Appel LJ, et al. (1999). Effect of dietary patterns on ambula-tory blood pressure: results from the Dietary Approaches to Stop Hypertension (DASH) Trial. DASH Collaborative Research Group. *Hypertension*, 34(3), 472-477.

[129] Conlin, P. R., Chow, D., Miller, E. R. I. I. I., et al. (2000). The effect of dietary patterns on blood pressure control in hypertensive patients: results from the Dietary Ap-proaches to Stop Hypertension (DASH) trial. Am J Hypertens.; , 13(9), 949-955.

[130] John, J. H., Ziebland, S., Yudkin, P., Roe, L. S., & Neil, H. A. (2002). Effects of fruit and vegetable consumption on plasma antioxidant concentrations and blood pres-sure: a randomised controlled trial. *Lancet*, 359(9322), 1969-1974.

[131] Parikh, A., Lipsitz, S. R., & Natarajan, S. (2009). Association between a DASH-like diet and mortality in adults with hypertension: findings from a population-based follow-up study. Am J Hypertens; , 22(4), 409-416.

[132] MRC/BHF(2002). Heart Protection Study of antioxidant vitamin supplementation in 20,536 high-risk individuals: a randomised placebo-controlled trial. Lancet; 6;, 360(9326), 23-33.

[133] Sesso HD, Buring JE, Christen WG, et al.(2008). Vitamins E and C in the prevention of cardiovascular disease in men: the Physicians' Health Study II randomized controlled trial. JAMA. 12;, 300(18), 2123-2133.

[134] Lee IM, Cook NR, Gaziano JM, et al.(2005). Vitamin E in the primary prevention of cardiovascular disease and cancer: the Women's Health Study: a randomized controlled trial. JAMA. 6; , 294(1), 56-65.

[135] Bjelakovic, G., Nikolova, D., Gluud, L. L., Simonetti, R. G., & Gluud, C. (2008). Antioxidant supplements for prevention of mortality in healthy participants and patients with various diseases. Cochrane Database Syst Rev. (2):, CD007176 EOF.

[136] Weinberg RB, VanderWerken BS, Anderson RA, Stegner JE, Thomas MJ.(2001). Prooxidant effect of vitamin E in cigarette smokers consuming a high polyunsaturated fat diet. Arterioscler Thromb Vasc Biol. , 21(6), 1029-1033.

[137] Salonen, J. T., Nyyssonen, K., Salonen, R., et al. (2000). Antioxidant Supplementation in Atherosclerosis Prevention (ASAP) study: a randomized trial of the effect of vitamins E and C on 3-year progression of carotid atherosclerosis. J Intern Med; , 248(5), 377-386.

[138] Münzel, T., Keaney, J. F., & Jr , . (2001). Are ACE-inhibitors a "magic bullet" against oxidative stress? *Circulation*, 104(13), 1571-1574.

[139] Ristow, M., Zarse, K., Oberbach, A., et al. (2009). Antioxidants prevent health-promoting effects of physical exercise in humans. Proc Natl Acad Sci U S A 26;, 106(21), 8665-8670.

[140] Stamler, J., Liu, K., Ruth, K. J., Pryer, J., & Greenland, P. (2002). Eight-year blood pressure change in middle-aged men: relationship to multiple nutrients. *Hypertension*, 39(5), 1000-1006.

[141] Siems, W., Sommerburg, O., Schild, L., Augustin, W., Langhans, C. D., & Wiswedel, I. (2002). Beta-carotene cleavage products induce oxidative stress in vitro by impairing mitochondrial respiration. FASEB J. , 16(10), 1289-1291.

[142] Upritchard JE, Sutherland WH, Mann JI. (2000). Effect of supplementation with tomato.juice, vitamin E, and vitamin C on LDL oxidation and products of inflammatory activity in type 2 diabetes. *Diabetes Care*, 23(6), 733-738.

[143] Engelhard, Y. N., Gazer, B., & Paran, E. (2006). Natural antioxidants from tomato extract reduce blood pressure in patients with grade-1 hypertension: a double-blind, placebo-controlled pilot study. Am Heart J. 151(1):100.

[144] Ried, K., Frank, O. R., & Stocks, N. P. (2009). Dark chocolate or tomato extract for prehypertension: a randomised controlled trial. BMC Complement Altern Med. 9:22.

[145] Chen, X., Touyz, R. M., Park, J. B., & Schiffrin, E. L. (2001). Antioxidant effects of vitamins C and E are associated with altered activation of vascular NADPH oxidase and superoxide dismutase in stroke-prone SHR. HypertensionPt 2):, 606 EOF-11 EOF.

[146] Ulker, S., Mc Keown, P. P., & Bayraktutan, U. (2003). Vitamins reverse endothelial dysfunction through regulation of eNOS and NAD(P)H oxidase activities. Hypertension. , 41(3), 534-539.

[147] Muhlhofer, A., Mrosek, S., Schlegel, B., et al. (2004). High-dose intravenous vitamin C is not associated with an increase of pro-oxidative biomarkers. Eur J Clin Nutr. , 58(8), 1151-1158.

[148] Fotherby, Williams. J. C., Forster, L. A., Craner, P., & Ferns, G. A. (2000). Effect of vitamin C on ambulatory blood pressure and plasma lipids in older persons. *Journal of Hypertension*, 18, 411-415.

[149] Mullan, Young. I. S., Fee, H., & Mc Cance, D. R. (2002). Ascorbic Acid reduces blood pressure and arterial stiffness in type 2 diabetes. *Hypertension*, 40(6), 804-809.

[150] Darko, D., Dornhorst, A., Kelly, F. J., Ritter, J. M., & Chowienczyk, P. J. (2002). Lack of effect of oral vitamin C on blood pressure, oxidative stress and endothelial function in Type II diabetes. Clin Sci (Lond) , 103(4), 339-344.

[151] McDermott JH.(2000). Antioxidant nutrients: current dietary recommendations and research update. J Am Pharm Assoc (Wash) , 40(6), 785-799.

[152] Upston, J. M., Witting, P. K., Brown, A. J., Stocker, R., Keaney, J. F., & Jr , . (2001). Effect of vitamin E on aortic lipid oxidation and intimal proliferation after arterial injury in cholesterol-fed rabbits. Free Radic Biol Med. 15;, 31(10), 1245-1253.

[153] Azzi, A., Ricciarelli, R., & Zingg, J. M. (2002). Non-antioxidant molecular functions of alpha-tocopherol (vitamin E) FEBS Lett. 2002 May 22;519(1-3):8-10.

[154] Yusuf, S., Dagenais, G., Pogue, J., Bosch, J., & Sleight, P. (2000). Vitamin E supplementation and cardiovascular events in high-risk patients. The Heart Outcomes Prevention Evaluation Study Investigators. N Engl J Med. , 342, 154-160.

[155] Miller, E. R. I. I. I., Pastor-Barriuso, R., Dalal, D., Riemersma, R. A., Appel, L. J., & Guallar, E. (2005). Meta-analysis: high-dosage vitamin E supplementation may increase all-cause mortality. Ann Intern Med. 2005 January 4;, 142(1), 37-46.

[156] Lonn, E., Bosch, J., Yusuf, S., et al. (2005). Effects of long-term vitamin E supplementation on cardiovascular events and cancer: a randomized controlled trial. JAMA. , 293(11), 1338-1347.

[157] Palumbo, G., Avanzini, F., Alli, C., et al. (2000). Effects of vitamin E on clinic and am-
       bulatory blood pressure in treated hypertensive patients. Collaborative Group of the
       Primary Prevention Project (PPP)--Hypertension study. Am J Hypertens. 13(5 Pt 1):,
       564 EOF-7 EOF.

[158] Ward NC, Wu JH, Clarke MW, et al.(2007). The effect of vitamin E on blood pressure
       in individuals with type 2 diabetes: a randomized, double-blind, placebo-controlled
       trial. J Hypertens. , 25(1), 227-234.

[159] Tiefenbacher CP.(2001). Tetrahydrobiopterin: a critical cofactor for eNOS and a strat-
       egy in the treatment of endothelial dysfunction? Am J Physiol Heart Circ Physiol. 280
       (6):HH2488., 2484.

[160] Govers, R., & Rabelink, T. J. (2001). Cellular regulation of endothelial nitric oxide
       synthase. Am J Physiol Renal Physiol. 280(2):FF206., 193.

[161] Katusic ZS(2001). Vascular endothelial dysfunction: does tetrahydrobiopterin play a
       role? Am J Physiol Heart Circ Physiol. 281(3):HH986., 981.

[162] Schlaich MP, Parnell MM, Ahlers BA, et al. (2004). Impaired L-arginine transport and
       endothelial function in hypertensive and genetically predisposed normotensive sub-
       jects. *Circulation*, 110(24), 3680-3686.

[163] Wang, D., Strandgaard, S., Iversen, J., & Wilcox, C. S. (2009). Asymmetric dimethylar-
       ginine, oxidative stress, and vascular nitric oxide synthase in essential hypertension.
       Am J Physiol Regul Integr Comp Physiol. 296(2):RR200., 195.

[164] Hidalgo Ponce Alejandro.(2007). Terapia Antioxidante. Foco en la microcirculación.
       Critical Care Medicine-Suppl., 35(9)

[165] Bevers, L. M., Braam, B., Post, J. A., et al. ((2006).) Tetrahydrobiopterin, but not L-
       arginine, decreases NO synthase uncoupling in cells expressing high levels of endo-
       thelial NO synthase. Hypertension. ., 47(1), 87-94.

[166] Matsuoka, H., Itoh, S., Kimoto, M., et al., Asymmetrical, dimethylarginine., an, en-
       dogenous., nitric, oxide., synthase, inhibitor., in, experimental., & hypertension, . Hy-
       pertension. (1997). January;29(1 Pt 2):242-247.

[167] Siani, A., Pagano, E., Iacone, R., Iacoviello, L., Scopacasa, F., & Strazzullo, . (2000).
       May;P. Blood pressure and metabolic changes during dietary L-arginine supplemen-
       tation in humans. Am J Hypertens. 13(5 Pt 1):547-551.

[168] Martina, V., Masha, A., Gigliardi, V. R., et al. (2008). Long-term N-acetylcysteine and
       L-arginine administration reduces endothelial activation and systolic blood pressure
       in hypertensive patients with type 2 diabetes. *Diabetes Care*, 31(5), 940-944.

[169] Kelly, Alexander. J. W., Dreyer, D., et al. (2001). Oral arginine improves blood pres-
       sure in renal transplant and hemodialysis patients. JPENJ Parenter Enteral Nutr. ,
       25(4), 194-202.

[170]   Gokce, N., & et, al. . (2004). L-arginine amd hypertension. J Nutr 134(10 Suppl) 2807S-28011S.

[171]   Loscalzo, J. (2003). Adverse effects of supplemental L-arginine in atherosclerosis: consequences of methylation stress in a complex catabolism? Arterioscler Thromb Vasc Biol. 1; , 23(1), 3-5.

[172]   Persky AM, Brazeau GA.(2001). Clinical pharmacology of the dietary supplement creatine monohydrate. Pharmacol Rev. , 53(2), 161-176.

[173]   Tyagi, N., Sedoris, K. C., Steed, M., Ovechkin, A. V., Moshal, K. S., & Tyagi, S. C. (2005). Mechanisms of homocysteine-induced oxidative stress. Am J Physiol Heart Circ Physiol. 289(6):HH2656., 2649.

[174]   Jahangir, E., Vita, J. A., Handy, D., et al., & (200, . (2009). The effect of l-arginine and creatine on vascular function and homocysteine metabolism. Vasc Med. , 14(3), 239-248.

[175]   Peters, U., Poole, C., & Arab, L. (2001). Does tea affect cardiovascular disease? a meta-analysis. Am J Epidemiol. 15; , 154(6), 495-503.

[176]   Bazzano, L. A., He, J., Ogden, L. G., et al. (2002). Fruit and vegetable intake and risk of cardiovascular disease in US adults: the first National Health and Nutrition Examination Survey Epidemiologic Follow-up Study. Am J Clin Nutr. , 76(1), 93-99.

[177]   Aviram, M., & Fuhrman, B. (2002). Wine flavonoids protect against LDL oxidation and atherosclerosis. Ann N Y Acad Sci. , 957, 146-161.

[178]   Lotito, S. B., & Frei, B. (2006). Consumption of flavonoid-rich foods and increased plasma antioxidant capacity in humans: cause, consequence, or epiphenomenon? Free Radic Biol Med. , 41(12), 1727-1746.

[179]   Aviram, M., & Dornfeld, L. (2001). Pomegranate juice consumption inhibits serum angiotensin converting enzyme activity and reduces systolic blood pressure. *Atherosclerosis*, 158(1), 195-198.

[180]   Aviram, M., Rosenblat, M., Gaitini, D., et al. (2004). Pomegranate juice consumption for 3 years by patients with carotid artery stenosis reduces common carotid intima-media thickness, blood pressure and LDL oxidation. Clin Nutr. , 23(3), 423-433.

[181]   Anter, E., Thomas, S. R., Schulz, E., Shapira, O. M., Vita, J. A., Keaney, J. F., & Jr , . (2004). Activation of eNOS by the MAP kinase in response to black tea polyphenols. J Biol Chem. 45:46637-46643., 38.

[182]   Diebolt, M., Bucher, B., & Andriantsitohaina, R. Wine polyphenols decrease blood pressure, improve NO vasodilatation, and induce gene expression. Hypertension(2001). August;, 38(2), 159-165.

[183]   Taubert, D., Berkels, R., Roesen, R., & Klaus, W. (2003). Chocolate and blood pressure in elderly individuals with isolated systolic hypertension. JAMA. , 290(8), 1029-1030.

[184] Grassi, D., Lippi, C., Necozione, S., Desideri, G., & Ferri, C. (2005). Short-term admin-
       istration of dark chocolate is followed by a significant increase in insulin sensitivity
       and a decrease in blood pressure in healthy persons. Am J Clin Nutr. , 81(3), 611-614.

[185] Taubert, D., Roesen, R., Lehmann, C., Jung, N., & Schomig, E. (2007). Effects of low
       habitual cocoa intake on blood pressure and bioactive nitric oxide: a randomized
       controlled trial. JAMA. , 298(1), 49-60.

[186] Grassi, D., Desideri, G., Necozione, S., et al. (2008). Blood pressure is reduced and in-
       sulin sensitivity increased in glucose-intolerant, hypertensive subjects after 15 days
       of consuming high-polyphenol dark chocolate. J Nutr. , 138(9), 1671-1676.

[187] Zilkens, R. R., Burke, V., Hodgson, J. M., Barden, A., Beilin, L. J., & Puddey, I. B.
       (2005). Red wine and beer elevate blood pressure in normotensive men. *Hypertension,*
       45(5), 874-879.

[188] Taubert, D., Roesen, R., & Schomig, E. (2007). Effect of cocoa and tea intake on blood
       pressure: a meta-analysis. Arch Intern Med. , 167(7), 626-634.

Chapter 7

Protective Effect of Silymarin on Liver Damage by Xenobiotics

José A. Morales-González, Evila Gayosso-Islas,
Cecilia Sánchez-Moreno, Carmen Valadez-Vega,
Ángel Morales-González, Jaime Esquivel-Soto,
Cesar Esquivel-Chirino, Manuel García-Luna y González-Rubio and
Eduardo Madrigal-Santillán

Additional information is available at the end of the chapter

1. Introduction

The liver is the vertebrates' largest internal organ. It weighs nearly 1.5 kg, is dark red in color, and is situated in the upper right quadrant of the abdominal cavity. Among the functions that it performs are the following: the metabolism of lipids and carbohydrates, and the synthesis of proteins, coagulation factors, and biliary salts. Eighty percent of the hepatic parenchyma is made up of hepatocytes, which are the cells mainly responsible for maintaining every function that the liver in its entirety requires to sustain the body's normal physiological functions in general. In addition to hepatocytes, the liver possesses other cells, such as the so-called Kupffer cells (hepatic macrophages), Ito cells, endothelial cells. The hepatocytes are disposed in the liver in groups denominated lobules, which have a central orifice comprised of the bile duct and by means of which the biliary salts are excreted. The anatomical loss of the structure of the hepatic lobule is considered a symptom of severe damage to the liver; it can be accompanied by partial or total loss of some physiological function, as in the case of alcohol-related hepatic cirrhosis. [23].

2. Hepatic regeneration

Liver regeneration is a fundamental response of the liver on encountering tissue damage. The complex interaction of factors that determine this response involves a stimulus (experimentally, a hepatectomy), gene expression, and the interaction of other factors that modulate the response. This proliferation depends on the hepatocytes, epithelial bile cells, Kupffer cells, and Ito cells. [24].

The mechanisms of hepatic growth have been studied in detail in experimental models. In the latter, regeneration is induced whether by tissue resection (partial hepatectomy) or by death of the hepatocytes (toxic damage). The principles that govern the growth of this organ in these systems also apply to clinical situations, such as, for example, fulminating liver failure, acute and chronic hepatitis, partial hepatectomy for treating liver cancer, or even in liver transplant donors. Evidence that there is a humoral growth factor of the hepatocyte has been observed in animal models and in patients with liver disease from the 1980s. [1, 13, 34, 10].

3. Ethanol

On being ingested, alcohol (also called ethanol) produces a series of biochemical reactions that lead to the affectation of numerous organs involving economy, having as the endpoint the development of hepatic diseases such as alcoholic hepatitis and cirrhosis. Despite that much is known about the physiopathological mechanisms that trigger ethanol within the organism, it has been observed that a sole mechanism of damage cannot fully explain all of the adverse effects that ethanol produces in the organism or in one organ in particular. [37, 30].

A factor that is referred as playing a central role in the many adverse effects that ethanol exerts on the organism and that has been the focus of attention of many researchers is the excessive generation of molecules called free radicals, which can produce a condition known as oxidative stress, which triggers diverse alterations in the cell's biochemical processes that can finally activate the mechanism of programmed cell death, also known as apoptosis. [28, 26, 17, 19, 25].

Of particular importance for the objective of this chapter is the focus on a particular class of free radicals that are oxygen derivatives, because these are the main chemical entities that are produced within the organism and that affect it in general.

4. Ethanol metabolism

Ethanol is absorbed rapidly in the gastrointestinal tract; the surface of greatest adsorption is the first portion of the small intestine with 70%; 20% is absorbed in the stomach, and the remainder, in the colon. Diverse factors can cause the increase in absorption speed, such as gastric emptying, ingestion without food, ethanol dilution (maximum absorption occurs at a 20% concentration), and carbonation. Under optimal conditions, 80-90% of the ingested dose is completely absorbed within 60 minutes. Similarly, there are factors that can delay ethanol absorption (from 2-6 hours), including high concentrations of the latter, the presence of food, the co-existence of gastrointestinal diseases, the administration of drugs, and individual variations [14, 37].

Once ethanol is absorbed, it is distributed to all of the tissues, being concentrated in greatest proportion in brain, blood, eye, and cerebrospinal fluid, crossing the feto-placentary and hematocephalic barrier [44]. Gender difference is a factor that modifies the distributed ethanol volume; this is due to its hydrosolubility and to that it is not distributed in body fats, which explains why in females this parameter is found diminished compared with males.

Ethanol is eliminated mainly (> 90%) by the liver through the enzymatic oxidation pathway; 5-10% is excreted without changes by the kidneys, lungs, and in sweat [14, 30]. The liver is the primary site of ethanol metabolism through the following three different enzymatic systems: Alcohol dehydrogenases (ADH); Microsomal ethanol oxidation system (MEOS); Catalase system.

5. Liver regeneration and ethanol

Ethanol is a well known hepatotoxic xenobiotic because hepatotoxicity has been well documented in humans as well as in animals. Although aspects concerning the pathogenesis of liver damage have been widely studied, it is known that liver regeneration restores the functional hepatic mass after hepatic damage caused by toxins. Suppression of the regenerating capacity of the liver by ethanol is the major factor of liver damage. [45]. Although the effects of acute or chronic administration of ethanol on the proliferative capacity of the liver to regenerate itself has been studied, the precise mechanism by which ethanol affects hepatocellular function and the regenerative process are poorly explained. [31, 29, 38].

Liver regeneration induced by partial hepatectomy in rats represents an ideal model of controlled hepatocellular growth. This surgical procedure has been sufficiently employed to study the factors than can be implicated in the growth of the liver. Endogenous signals have been described to control hepatic regeneration. The first marker of DNA synthesis in partially hepatectomized rats (70%) occurring normally 24-28 hours postsurgery comprises an enormous action of growth factors and cytokines affecting expression of the gene of the hepatocytes, associated with initiation of the cell cycle. [2]. It has indicated that the hepatocytes enter into a state denominated "priming" to thus begin replication and response to growth factors, that is, which range from the quiescent to the G 1 phase of the cell cycle. The progression of hepatic cells requires the activation of cyclin-dependent kinases that are regulated by cyclins and cyclin-dependent kinase inhibitors. [39]. It has also been demonstrated that a dose of ethanol importantly diminishes the specific activity of two enzymes related with the metabolism of DNA synthesis, which are thymidine kinase and thymidylate synthetase. [47, 11, 32].

6. Free radicals

Free radicals are the result of the organism's own physiological processes, such as the metabolism of food, respiration, exercise, or even those generated by environmental factors, such as contamination, tobacco, or by drugs, chemical additives, etc. Free radicals (FR) are atoms or groups of atoms that in their atomic structure present one or more unpaired electrons (odd in number) in the outer orbit. This spatial configuration generates in the molecule distinct physical and chemical properties such as heightened reactivity and diminished lifetime, respectively. [5, 27].

This instability confers on these physical avidity for the uptake of an electron of any other molecule in its ambit (stable molecules), causing the affected structure to remain unstable with the purpose of reaching its electrochemical stability. Once the free radical has achieved trapping the electron that it requires for pairing with its free electron, the stable molecule that cedes the latter to it in turn becomes a free radical, due to its remaining with an unpaired electron, this initiating a true chain reaction that destroys our cells. [4, 7].

In aerobic cells, there are diverse pathways that lead to the production of Oxygen-derived free radicals (OFR). The main sources are enzymes associated with the metabolism of arachidonic acid, such as cycloxygenase, lipoxygenase, and cytochrome P-450. The presence and ubiquity of enzymes (superoxide dismutase, catalase, and peroxidase) that eliminate secondary products in a univalent pathway in aerobic cells suggest that the superoxide anions and hydrogen peroxide are important secondary products of oxidative metabolism. [40, 7]. Reactive oxygen species (ROS) can damage macromolecules such as DNA, carbohydrates, and proteins. These cytotoxic oxygen species can be classified as two types:

1.   the free radicals, such as the superoxide radical (O_2) and the hydroxyl radical ($^{.}OH$), and

2.   non-radical oxygen species, such as hydrogen peroxide (H_2O_2), the oxygen singlette (O_1), which is a very toxic species, peroxynitrite (ONOO), and Hypochlorous acid (HOCL).

The instable radicals attack cell components, causing damage to the lipids, proteins, and the DNA, which can trigger a chain of events that result in cellular damage. [7, 21]. These reductive processes are accelerated by the presence of trace metals such as iron (Fe) and copper (Cu) and of specific enzymes such as monoxygenases and certain oxidases. [7, 21]

7. Oxidative stress

In 1954, an Argentine researcher, Rebeca Gerschman, suggested for the first time that FR were toxic agents and generators of disease. [12].

Due to the atomic instability of FR, the latter collide with a biomolecule and subtract an electron, oxidating it, losing in this manner its specific function in the cell. If lipids are involved (polyunsaturated fatty acids), the structures rich in these are damaged, such as the cell membranes and the lipoproteins. In the former, the permeability is altered, leading to edema and cell death, and in the latter, to oxygenation of the Low-density lipoproteins (LDL) and genesis of the atheromatous plaque. The characteristics of lipid oxygenation by FR involve a chain reaction in which the fatty acid, on being oxygenated, becomes a fatty acid radical with the capacity of oxidizing another, neighboring molecule. This process is known as lipid peroxidation and it generates numerous subproducts, many of these, such as Malondialdehyde (MDA), whose determination in tissues, plasma, or urine is one of the methods for evaluating oxidative stress. In the case of proteins, these preferentially oxidize the amino acids (phenylalanine, tyrosine, triptophan, histidine, and methionine), and consequently form peptide chain overlapping, protein fragmentation, and the formation of carbonyl groups, and these impede the normal development of their functions (ionic membrane transporters, receptors, and cellular messengers), enzymes that regulate the cell's metabolism, etc.). [7, 33].

8. Liver regeneration, ethanol, and free radicals

While ROS and FR are generated during ethanol metabolism, causing oxidative stress and lipoperoxidation in the liver, they can also form a significant pathway of damage to the regenerative process of the hepatocyte. In this process, ethanol-induced FR and the generation of ROS involve the mitochondria, the microsomal cytochrome P450 2E1, the iron (FE) ion, and less frequently, peroxisomes, cytosolic xanthines, and aldehyde oxidases, to regulate cellular proliferation, acting as direct or indirect factors. [37, 2].

In general, ROS-derived FR intervene in the persistent bombardment of molecules by reactive oxygen radicals, thus maintaining redox homeostasis, in such a manner that during liver regeneration, these can modify the metabolic response necessary for carrying out cellular mitosis in the hepatocyte. While FR are generated and utilized by the cells as neutrophils, monocytes, macrophages, eosinophils, and fibroblasts for eliminating foreign organisms or toxic substances, the increase of FR due to exposure to ethanol leads to cellular deterioration that in turn produces hepatic alterations, with an unfavorable influence on cell proliferative action. [25].

9. Antioxidants

Halliwell defines an antioxidant as all substances that on being found present at low concentrations with respect to those of an oxidizable substrate (biomolecule), delays or prevents the oxidation of this substrate. The antioxidant, on colliding with FR, is ceded to an electron, in turn oxidizing itself and transforming itself into a non-toxic, weak FR and, in some cases such as with vitamin E, it can regenerate itself into its primitive state due to the action of other antioxidants. Not all antioxidants act in this way: the so-called enzymatic antioxidants catalyze or accelerate chemical reactions that utilize substrates that in turn react with FR. Of the numerous classifications of antioxidants, it is recommended to adopt that which divides these into the following: exogenes or antioxidants that enter through the alimentary chain, and endogenes that are synthesized by the cell. Each antioxidant possesses an affinity for a determined FR or for several. Vitamin E, beta-carotene, and lycopene act within the liposoluble medium of the cell and their absorption and transport are found to be very much linked with that of the lipids. Vitamin E is considered the most important protector of lipid molecules. [27].

Life in the presence of molecular oxygen requires the possession of a multiple battery of defenses against the diverse oxygen FR, which on the one hand tend to impede their formation and on the other, neutralize them once they are formed. These defenses exert an effect at five levels [7, 21, 33]:

9.1. First level

This consists of editing univalent oxygen reduction through enzymatic systems capable of effecting consecutive tetravalent reduction without releasing the partially reduced intermediaries; this is achieved with great effectiveness by the cytochrome-oxidase system of the mitochondrial respiratory chain, which is responsible for more than 90% of oxygen reduction in the human organism.

9.2. Second level

This is constituted of enzymes specialized in the uptake of the superoxide anion radical (O_2 –). These are Superoxide dismutase (SOD), the methaloenzyme that catalyzes the dismutation of the superoxide anion radical to provide molecular oxygen and hydrogen peroxide, with such great effectiveness that it approaches the theoretical limit of diffusion. In the cells of the eukaryotic organisms, there are two of these: one is cytoplasmatic, and the other is mitochondrial. SOD was described by Fridovich in 1975.

9.3. Third level

This is conferred by a group of specialized enzymes on neutralizing hydrogen peroxide. Among these is catalase, which is found in the peroxisomes and which catalyzes the dismutation reaction.

Also in mammals, glutathione peroxidase (a cytoplasmic enzyme that contains selenium) is the most important.

9.4. Fourth level

Here the hydroxyl radical produced in the Haber-Weiss cycle can neutralized by vitamin E or alpha-tocopherol, which is an effective antioxidant and that due to its hydrophobicity is found in biological membranes in which its protection is particularly important. In addition, vitamin C or ascorbic acid is a reducer agent or electron donor and reacts rapidly with the OH– radical and with the superoxide anion.

9.5. Fifth level

Once the molecular damage is produced, there is a fifth level of defense that consists of repair. It has been demonstrated that FR were capable of causing breaks in the DNA chain and even of inducing mutagenesis, but there are enzymatic repair mechanisms that permit reestablishment of genetic information.

10. Antioxidants and their role in hepatoprotection

The term antioxidant was originally utilized to refer specifically to a chemical product that prevented the consumption of oxygen [6]; thus, antioxidants are defined as molecules whose function is to delay or prevent the oxygenation of other molecules. The importance of antioxidants lies in their mission to end oxidation reactions that are found in the process and to impede their generating new oxidation reactions on acting in a type of sacrifice on oxidating themselves. There are endogenous and exogenous antioxidants in nature. Some of the best-known exogenous antioxidant substances are the following: β-carotene (pro-vitamin A); retinol (vitamin (A); ascorbic acid (vitamin C); α-tocopherol (vitamin E); oligoelements such as selenium; amino acids such as glycine, and flavonoids such as *silymarin*, among other organic compounds [46, 36].

Historically, it is known that the first investigations on the role that antioxidants play in Biology were centered on their intervention in preventing the oxidation of unsaturated fats, which is the main cause of rancidity in food. However, it was the identification of vitamins A, C, and E as antioxidant substances that revolutionized the study area of antioxidants and that led to elucidating the importance of these substances in the defense system of live organisms. [36].

Due to their solubilizing nature, antioxidant compounds have been divided into hydrophilics (phenolic compounds and vitamin C) and lipophilics (carotenoids and vitamin E). The antioxidant capacity of phenolic compounds is due principally to their redox properties, which allow them to act as reducing agents, hydrogen and electron donors, and individual oxygen inhibitors, while vitamin C's antioxidant action is due to its possessing two free electrons that can be taken up by Free radicals (FR), as well as by other Reactive oxygen species (ROS), which lack an electron in their molecular structure. Carotenoids are deactivators of electronically excited sensitizing molecules, which are involved in the generation of radicals and individual oxygen, and the antioxidant activity of vitamin A is characterized by hydrogen donation, avoiding chain reactions. [7, 21, 33].

The antioxidant defense system is composed of a group of substances that, on being present at low concentrations with respect to the oxidizable substrate, delay or significantly prevent oxygenation of the latter. Given that FR such as ROS are inevitably produced constantly during metabolic processes, in general it may be considered as an oxidizable substrate to nearly all organic or inorganic molecules that are found in living cells, such as proteins, lipids, carbohydrates, and DNA molecules. Antioxidants impede other molecules from binding to oxygen on reacting or interacting more rapidly with FR and ROS than with the remainder of molecules that are present in the microenvironment in which they are found (plasma membrane, cytosol, the nucleus, or Extracellular fluid [ECF]). Antioxidant action is one of the sacrifices of its own molecular integrity in order to avoid alterations in the remainder of vitally functioning or more important molecules. In the case of the exogenic antioxidants, replacement through consumption in the diet is of highest importance, because these act as suicide molecules on encountering FR, as previously mentioned. [7, 21, 33].

This is the reason that, for several years, diverse researchers have been carrying out experimental studies that demonstrate the importance of the role of antioxidants in protection and/or hepatic regeneration in animals. Thus, in this chapter, the principal antioxidants will be described that play an important role in the regeneration of hepatic cells and in the prevention of damage deriving from alcohol.

11. Flavonoids

Flavonoids are compounds that make up part of the polyphenols and are also considered essentials nutrients. Their basic chemical structure consists of two benzene rings bound by means of a three-atom heterocyclic carbon chain. Oxidation of the structure gives rise to several families of flavonoids (flavons, flavonols, flavanons, anthocyanins, flavanols, and isoflavons), and the chemical modifications that each family can undergo give rise to >5,000 compounds identified by their particular properties. [16].

Flavonoid digestion, absorption, and metabolism have common pathways with small differences, such as, for example, unconjugated/non-conjugated flavonoids can be absorbed at the stomach level, while conjugated flavonoids are digested and absorbed at the intestinal level by extracellular enzymes on the enterocyte brush border. After absorption, flavonoids are conjugated by methylation, sukfonation, ands glucoronidation reactions due to their biological activity, such as facilitating their excretion by biliary or urinary route. The conjugation type the site where this occurs determine that metabolite's biological action, together with the protein binding for its circulation and interaction with cellular membranes and lipoproteins. Flavonoid metabolites (conjugated or not) penetrate the tissues in which they possess some function (mainly antioxidant), or are metabolized. [27].

On the other hand, the flavonoids possess implications in health; in recent years, the properties of these compounds have been studied in relation to diverse pathologies. In diabetes, these compounds present regulation of glycemia through diverse mechanisms that include the inhibition of some enzymes such as α-glucosidase, glucose 6 phosphatase, and phosphorylated glycogen. The flavonoids possess other characteristics such as the trapping of molecules of glioxal and methyl-glioxal molecules, which propitiate the formation of advanced final products of glycosylation that are found to be directly related with micro- and macrovascular complications. They also regulate the rise or fall of transporter proteins; the structure of some flavonoids appears to have important participation with regard to the studied benefits. [16].

More research is needed because great majority of the former has been conducted in animals, to determine effects and dosage. Flavonoids in the menopause result in controversial effects due to the population type studied, that is, Asiatic, absorption, metabolism, the binding of isoflavones to estrogen receptors, etc.; however, they appear to possess a beneficial effect in terms of the prevention of certain types of cancer and osteoporosis. [16].

The flavonoids absorb Ultraviolet light (UV) from the sun and possess direct and indirect antioxidant effects (through the induction of cytoprotector proteins). Topical application (on the human skin) of the polyphenolic fraction of green tea protects against immunosuppression and inhibits the erythema and the formation of pyrimidine dimers in DNA caused by UV. On representing one of the most important lifestyle factors, alimentation can importantly affect the incidence and initiation of cardiovascular or neurodegenerative diseases. The cardioprotector effect of flavonoids is based on reducing oxidation and blood concentrations of the binding of cholesterol to Low-density lipoproteins (LDL); flavonoids reduce endothelial dysfunction and blood pressure and increase the HDL-bound cholesterol concentration. Flavonoids possess a neuroprotector effect because they protect the neurons from oxidative stress by means of induction of antioxidant defenses, modulation of signaling cascades, mitochondrial interactions, apoptotic processes, or by synthesis/degradation of the β-amyloid peptide. The potential effect of flavonoids as neuroprotectors is due to three main factors: they prevent neurodegeneration; inhibit neuroinflammation, and reduce the diminution of age-related cognitive functions. [16].

In cancer, the flavonoids have been classified as chemopreventive, as blockers as well as inhibitors, given their functions in carcinogenesis, in which they modulate transduction signaling in cellular proliferation and angiogenesis, modulate enzymes for the metabolic activation of procarcinogens and the detoxification of carcinogens, and modulate enzymes in the biosynthesis of anti-oxidant-pro-oxidant estrogen activity estrogen (promoting oxidative homeostasis, rendering its antioxidative capacity as a contribution to antineoplastic as well as preventive as well as therapeutic activity due to inhibiting the activation of mitogenic kinases and transduction factors, while pro-oxidative activity increases the cell damage that promotes detention of the cell cycle and apoptosis). In obesity, the flavonoids have been identified as reducer factors of fat mass and as inhibitors of fat mass deposition and catabolic activity. [16].

The procyanidins and proanthocyanidins have demonstrated, in human population, to diminish visceral fatty mass (depending on the dose) with an associated increase of adiponectin. This diminution is linked with the malabsorption of carbohydrates and lipids due to enzyme inhibition. It has been observed that the procyanidins increase β-oxidation and inhibit the expression of genes that promote the synthesis of fatty acids. Epigallocatechin gallate can increase energy expenditure and lipid oxidation in humans; it is thought that this is possible because of the increase of thermogenesis and the inhibition of the activity of the lipase, as well as, according to studies *in vitro*, the inhibition of lipogenesis and apoptosis of the adipocytes. Catechins that alter the deposition of adipose tissue related with diminution of the respiratory co-efficient and greater oxygen consumption, and thermogenesis induced by the sympathetic nervous system. Phytoestrogens can improve obesity and its alterations on diminishing insulin resistance, thus lipogenesis, as well as inhibition of the mechanisms for cell differentiation and proliferation. The study of flavonoids and their effects on the prevention and treatment of obesity is a widespread, yet incomplete research field. [16].

The metabolism of phytoestrogens and their maximum concentration in serum presents great variability, depending on genetic differences and estrogen exposure in early life stages. [16].

12. Silimarina (*silybum marianum*)

Silymarin is a compound of natural origin extracted from the *Silybum marianum* plant, popularly known as St. Mary's thistle, whose active ingredients are flavonoids such as silybin, silydianin, and silycristin. This compound has attracted attention because of its possessing antifibrogenic properties, which have permitted it to be studied for its very promising actions in experimental hepatic damage. In general, it possesses functions such as its antioxidant one, and it can diminish hepatic damage because of its cytoprotection as well as due to its inhibition of Kupffer cell function. [41].

Silymarin, derived from the milk thistle plant named *Silybum marianum*, has been used since time past as a natural remedy for combating liver diseases. *Silymarin* and its active constituents (silybinin, silycristin, and silydianine, among others), have been classified as uptakers of free radicals and inhibitors of lipoperoxidation; some studies also suggest that that they increase the synthesis of hepatocytes, diminish the activity of tumor promoters, stabilize mastocyte cells, and act as iron chelates. [8].

Silybum marianum belongs to the Aster family (Asteraceae or Compositae), which includes daisies and thistles. The milk thistle is distributed widely throughout Europe, was the first plant that appeared in North America to the European colonizers, and is at present established in the South of the U.S., California, and South America. [22].

The name milk thistle is derived from the characteristics of its thorny leaves with white veins, which, according to the legend, were carried by the Virgin Mary. Its name *Cardo lechoso* derives from the same tradition. The mature plant has large flowers, of a brilliant purple color, and abundant thorns of significant appearance. The milk thistle grows in places where exposure to the sun is abundant. [15].

Extracts of the milk thistle have been used as medical remedies from ancestral Greece, when Dioscorides, a Greek herbalist, wrote that the seeds of the milk thistle could cure the bite of a poisonous snake. Pliny noted that the mixture of the juice of the plant and its honey were excellent for bile tract disorders. [9]. In 1596, Gerard mentioned *Silybum marianum* as a major remedy against melancholy or black bile. The milk thistle was sold for treating liver diseases. In the 1960s, observed that milk thistle was an excellent remedy for cleaning obstructions of the liver and spleen, notwithstanding that infusions of the fresh roots and seeds were effective for counteracting jaundice.

The main active agent of the milk thistle is *silymarin*, a mixture of flavonolignans, silydia-nine, silycristin, and silybinin, the latter the most biologically active extract; the flavonoids appear to be activated as trappers of free radicals and as plasmatic membrane stabilizers. Concentrations of *silymarin* are localized in the fruit of the plant, as well as in the seeds and leaves, from which *silymarin* is extracted with 95%-proof ethanol, achieving a brilliant yellow liquid. The term flavonoid is derived from *flavus*, which denotes yellow. [20, 16].

The standardized extract of *silymarin* contains 70% *silymarin*. Pharmacokinetic studies have shown that there is rapid absorption of silybinin into the bloodstream after an oral dose. Steady-state plasma concentrations are reached after 2 hours and the elimination half-life is 6 hours. [Lorenz et al., 1984, 3]. From 3-8% of an oral dose is excreted in the urine and from 20-40% is recovered in the bile as glucuronide and sulfate. [42].

Silybinin works as an antioxidant, reacting rapidly with oxygen free radicals as demonstrated *in vitro* with hydroxyl anions and hypochlorous acid. Reported activities include the inhibition of hepatocyte lipoperoxidation, the microsomal membrane in rats, and protection against genomic damage through the suppression of hydrogen peroxide, superoxide anions, and lipoxygenase. It is thought that silybinin also increases the synthesis of the proteins of the hepatocyte through stimulation of the activity of the ribosomal RNA (rRNA) polymerase. In addition, silybinin diminishes hepatic and mitochondrial oxidation induced by an iron overcharge and acts as an iron chelate. [16].

13. Antioxidant and hepatoprotector action

Silymarin is an active principle that possesses hepatoprotector and regenerative action; its mechanism of action derives from its capacity to counterarrest the action of FR, which are formed due to the action of toxins that damage the cell membranes (lipid peroxidation), competitive inhibition through external cell membrane modification of hepatocytes; it forms a complex that impedes the entrance of toxins into the interior of liver cells and, on the other hand, metabolically stimulates hepatic cells, in addition to activating RNA biosythesis of the ribosomes, stimulating protein formation. In a study published by [41], the authors observed that *silymarin*'s protector effect on hepatic cells in rats when they employed this as a comparison factor on measuring liver weight/animal weight % (hepatomegaly), their values always being less that those of other groups administered with other possibly antioxidant substances; no significant difference was observed between the *silymarin* group and the *silymarin*-alcohol group, thus demonstrating the protection of *silymarin*. On the other hand, *silymarin* diminishes Kupffer cell activity and the production of glutathione, also inhibiting its oxidation. Participation has also been shown in the increase of protein synthesis in the hepatocyte on stimulating polymerase I RNA activity. *Silymarin* reduces collagen accumulation by 30% in biliary fibrosis induced in rat. An assay in humans reported a slight increase in the survival of persons with cirrhotic alcoholism compared with untreated controls [2].

Silymarin is a flavonoid derived from the *Silybum marianum* plant that has been employed for some 2,000 years for the treatment of liver diseases. At present, its use as an alternative drug has extended throughout Europe and the U.S. *Silymarin* acts as a hepatoprotector due to its antioxidant effect, which has been observed to inhibit liver damage due to the releasing of the substances of free radicals, such as ethanol, acetaminophen, and Carbon tetrachloride (CCL_4), in addition to increasing the activity of SOD and glutathione. As a uptaker of free radicals, *silymarin* can inhibit the lipid peroxidation cascade in the cell membranes. The hepatoprotector effect of this flavonoid also can be explained by an anti-inflammatory effect, in which it has been observed that *silymarin* acts on the functions of the Kupffer cells. Inhibition also has been reported in the activation of the Nuclear kappa-Beta [NK-B] transcription factor. [2, 16, 7, 21, 33].

14. *Silymarin* and Exercise

During physical activity, oxygen consumption increases, which produces oxidative stress that leads to the generation of free radicals, which are highly toxic for the cell, because these interact with organic molecules susceptible to being oxidized, such as unsaturated fatty acids, which causes lipoperoxidation. To avoid this damage by FR, there are the following antioxidant systems: Superoxide-dismutase (SOD), and Catalase (CAT), in addition to other protector substances such as vitamins A, C, and E and the flavonoids, which trap free radicals. (unpublished data)

In experiments conducted by our research group on groups of rats that were submitted to daily aerobic exercise in a physical-activity cage for 20 minutes during 4 weeks (5 days/week) and on another group of rats submitted to physical activity plus administration of *silymarin* (200 mg/kg of weight) prior to exercise, with daily quantification of physical performance and at the at the end of the experiment, quantification of DNA in serum and of SOD and CAT activity in liver. We found that in the group with physical activity, MDA increased 134% (in serum) and 123% (in liver) vs. control rats. In the group with exercise plus *silymarin*, MDA returned to normality (in serum and in liver). Catalase activity increases during exercise (118%) and with exercise plus *silymarin* (137%). SOD activity exhibited no modifications in any treatment. Finally, we found an increase of physical activity in the group administered *silymarin* (27%) in comparison with the group in which no *silymarin* was administered. (unpublished data)

A protector effect was found of *silymarin* during exercise, because it diminishes MDA levels in serum as well as in liver, which translates into diminution of the production of free radicals, causing as a consequence less cellular damage, which in turn leads to an increase in physical performance.

15. Conclusions

The process of the induction of oxidative stress generated in the liver due to the presence of ethanol implies the conjugation of various factors. The role that these factors play in the development of oxidative stress depends in part on whether acute or chronic intoxication is involved. The factors that contribute to the development of oxidative stress imply disequilibrium among pro- and antioxidant factors. It can occur that oxidative stress develops if the xenobiotic increases the pro-oxidant factors (the generation of Oxygen-generated free radicals [OFR]) or decreases intracellular antioxidant factors. In whichever of the two cases, the general result is important damage to the hepatocyte that can lead to general damage to the DNA that, in turn, can comprise a determining factor in the induction of the apoptotic system (programmed cell death) of the cell, thus accelerating its death and destruction.

The study of the factors that determine the increase in the generation of OFR in the liver, originating due to acute or chronic intoxication with a xenobiotic, is of great importance because it will allow diminishing the damage that these reactive species produce within the hepatocyte. On the other hand, despite that at present much is known concerning the physiopathological mechanisms of ethanol ingestion-related liver damage and the role that the production of oxygen-generated free radicals plays in these processes, the exact extent of this damage, as well as how to prevent it, remains unknown with precision. There is evidence obtained from laboratory models that the ingestion of natural antioxidants, such as vitamins A, C, and E, oligoelements (selenium), amino acids (glycine), and principally flavonoids, such as *silymarin*, can in the future be a potential treatment for all persons who present hepatic alterations. However, beyond the remedy, the cooperation of the patient is required to regulate his/her ethanol consumption; as long as this does not take place, taking antioxidant vitamins can be considered within the regular therapy of a patient with alcoholism, taking care above all that this supplement does not reach toxic concentrations, in particular in the case of the vitamin that possess the tendency to accumulate in the liver.

The use of novel experimental procedures that determine the degree of damage caused by xenobiotics, and in particular by free radicals, is of great importance in the management of diseases caused by this type of substance, especially if they damage the liver, because this organ comprises a vital part of our organism on having in its charge the metabolic support of the latter.

Author details

José A. Morales-González[1], Evila Gayosso-Islas[1], Cecilia Sánchez-Moreno[1],
Carmen Valadez-Vega[1], Ángel Morales-González[2], Jaime Esquivel-Soto[3],
Cesar Esquivel-Chirino[3], Manuel García-Luna y González-Rubio[3] and
Eduardo Madrigal-Santillán[1]

1 Instituto de Ciencias de la Salud, UAEH, México

2 Escuela Superior de Computo, IPN, México

3 Facultad de Odontología, UNAM, México

References

[1] Arakaki, N., Kawatani, S., Nakamura, O., & Ohnishi, T. (1995). Evidence for the presence of an inactive precursor of human hepatocyte growth factor in plasma and sera of patients with liver diseases. *Hepatology*. Vol. 22, pp. 1728-1734.

[2] Baptista, P. (2012). *Liver Regeneration*. Ed. Intech, Croatia, 252 pp.

[3] Barzaghi, N., Crema, F., Gatti, G., et al. (1990). Pharmacokinetic studies in IdB1016, a silybin-phosphatidylcholine complex, in healthy human subjects. *Em J Drug Metab Pharmacokinet*. Vol. 15, pp.333-338.

[4] Bergendi, LL., Benes, Z., & Durackiova y, M. F. (1999). Chemistry, physiology and pathology of free radicals. *Life Sciences*. Vol.64, pp. 1865-1874.

[5] Brunk, U., & Cadenas, E. (1988). The potential intermediate role of lysosomes in oxygen free radical pathology. *Review article*. Vol. 96, pp. 3-13.

[6] Burneo-Palacios, ZL. (2009). Determinación del contenido de compuestos fenólicos totales y actividad antioxidante de los extractos totales de doce especies vegetales nativas del sur del Ecuador (Tesis) Loja, Ecuador: Universidad Técnica Particular de Loja. Disponible en: http://es.scribd.com/doc/43393190/TESIS-ANTIOXIDANTES

[7] Camacho-Luis, A., Mendoza-Pérez, JA. (2009). La naturaleza efímera de los radicales libres. Química y bioquímica de los radicales libres. In *Los antioxidantes y las enfermedades crónico degenerativas*. Morales-González, JA., Fernández-Sánchez, AM., Bautista-Ávila, M., Vargas-Mendoza, N., Madrigal-Santillán, EO (ed). Ed. UAEH, Pachuca, Hidalgo, México, pp. 27-76.

[8] Flora, K., Hahn, M., Rosen, H., & Benner, K. (1998). Milk thistle (Silybum marianum) for the therapy of liver disease. *American Journal of Gastroenterology*. Vol 93, pp. 139-143.

[9] Foster, S. (1991). Milk thistle: Silybum marianum. Austin, TX: AmericanBotanical Council, No. 305.

[10] Fujiwara, K., Nagoshi, S., Ohno, A., Hirata, K., Ohta, Y., & Mochida, S. (1993). Stimulation of liver growth factor by exogenous human hepatocyte growth factor in normal and partially hepatectomized rats. *Hepatology*. Vol. 18, pp. 1443-1449.

[11] George, D., Liatsos, MD., et al. (2003). Effect of Acute Ethanol Exposure on Hepatic Stimulator Substance (HSS) leves During Liver Regeneration. *Digestive Diseases and Sciences*. Vol 48, pp. 1929-1938.

[12] Gerschman, R. (1954). Oxigen poisoning and X-Irradiation. A mechanism in common. *Science*. Vol. 119, pp. 623-626.

[13] Ghoda, E., Tsubouchi, H., Nakayama, H., & Hirono, S. (1988). Purification and partial characterization of hepatocyte growth factor from plasma of a patient with fulminant hepatitis failure. *J Clin Invest*. Vol. 81, pp. 414-419.

[14] Goldfrank, L., Flomenbaum, N., & Lewin, N. (2002). Goldrank's Toxicology Emergencies. 7th. Ed. McGraw-Hill, USA, pp. 952-962.

[15] Greive M. (1981). A modern herbal, vol. 2. New York: Dover Publications.

[16] Guillén-López, S., Álvarez-Salas, E., & Ochoa-Ortiz, E. (2009). Antioxidantes en el tratamiento de las enfermedades: flavonoides. In Los antioxidantes y las enfermedades crónico degenerativas. Morales-González, JA., Fernández-Sánchez, AM., Bautista-Ávila, M., Vargas-Mendoza, N., Madrigal-Santillán, EO (ed). Ed. UAEH, Pachuca, Hidalgo, México, pp. 593-615.

[17] Gutiérrez Salinas, J., & Morales-González (2004). Producción de radicales libres derivados del oxígeno y el daño al hepatocito. Revista de Medicina Interna de México. Vol. 20, pp. 287-295.

[18] Gutierrez-Salinas, J. (2007). Daño al hígado por radicales libres derivados del oxígeno. In Alcohol, alcoholismo y cirrosis. Un enfoque multidisciplinario. Morales-González, JA (ed). Ed. UAEH, Pachuca, Hidalgo, México, pp. 97-109.

[19] Gutiérrez-Salinas, J., & Morales-González, JA. (2006). La ingesta de fluoruro de sodio produce estrés oxidativo en la mucosa bucal de la rata. Revista Mexicana de Ciencias Farmacéuticas. Vol. 37, pp. 11-22.

[20] Harnisch, G., & Stolze, H. (1983). Silybum marianum: Mariendistel. In: BewaehrtePflanzendrogen in Wissenschaft und Medizin. Notamed Verlag, pp. 203-215.

[21] Hernández-Ceruelos, MCA., Sánchez Gutiérrez, M., Fragoso Antonio, S., Salas Guzmán, D., Morales-González, JA., Madrigal Santillán, EO. (2009). Quimioprevención de fitoquímicos. In Los antioxidantes y las enfermedades crónico degenerativas. Morales-González, JA., Fernández-Sánchez, AM., Bautista-Ávila, M., Vargas-Mendoza, N., Madrigal-Santillán, EO (ed). Ed. UAEH, Pachuca, Hidalgo, México, pp. 77-89.

[22] Hobbs, C. (1992). Milk thistle: The liver herb. Capitola, CA: Botanical Press.

[23] Koolman, J. Rohm. (2005). Bioquímica, Texto y Atlas. 3ra edición, panamericana. pp. 306.

[24] Michalopoulos, GK., & DeFrances, MC. (1997). Liver regeneration. Science Vol. 276, pp. 60-66.

[25] Morales-González, JA., Barajas-Esparza, L., Valadez-Vega, C., Madrigal-Santillán, E., Esquivel-Soto, J., Esquivel-Chirino, C., Téllez-López, AM., López-Orozco, M., & Zúñiga-Pérez, C. (2012). The Protective Effect of Antioxidants in Alcohol Liver Damage In: Liver Regeneration. Baptista, P. (ed). Ed. Intech, Croacia, pp. 89-120.

[26] Morales-González, JA., Bueno-Cardoso, A., Marichi-Rodríguez, F., & Gutiérrez-Salinas, J. (2004a). Programmed cell death (apoptosis): the regulating mechanisms of cellular proliferation. Arch Neurocien. Vol. 9, pp. 124-132.

[27] Morales-González, JA., Fernández-Sánchez, Bautista-Ávila, M., Vargas-Mendoza, N., & Madrigal-Santillán, EO. (2009). *Los antioxidantes y las enfermedades crónico degenerativas.* Ed. UAEH, Pachuca, Hidalgo, México, 751 pp.

[28] Morales-González, JA., Gutiérrez-Salinas, J., & Hernández-Muñoz, R. (1998). Pharmacokinetics of the ethanol bioavalability in the regenerating rat liver induced by partial hepatectomy. *Alcoholism Clinical and Expimental Research.* Vol. 22, pp. 1557-1563.

[29] Morales-González, JA., Gutierrez-Salinas, J., & Piña, E. (2004b). Release of Mitochondrial Rather than Cytosolic Enzymes during Liver Regeneration in Ethanol-Intoxicated Rats. *Archives of Medical Research.* Vol. 35, pp. 263-270.

[30] Morales-González, JA., Gutiérrez-Salinas, J., Arellano-Piña, G., Rojas-López, M., & Romero-Pérez, L. (1998). El metabolismo hepático del etanol y su contribución a la enfermedad hepática por etanol. *Revista de Medicina Interna de México.* Vol. 14, pp. 180-185.

[31] Morales-González, JA., Gutiérrez-Salinas, J., Yánez, L., Villagómez, C., Badillo, J. & Hernández, R. (1999). Morphological and biochemical effects of a low ethanol dose on rat liver regeneration. Role of route and timing of administration. *Digestive Diseases and Sciences.* Vol. 44, No. 10 (October), pp. 1963-1974.

[32] Morales-González, JA., Jiménez, L., Gutiérrez-Salinas, J., Sepúlveda, J., Leija, A. & Hernández, R. (2001). Effects of Etanol Administration on Hepatocellular Ultraestructure of Regenerating Liver Induced by Partial Hepatectomy. *Digestive Diseases and Sciences.* Vol. 46, No. 2 (February), pp. 360–369.

[33] Muñoz Sánchez, JL. (2009). Defensas antioxidantes endógenas. In *Los antioxidantes y las enfermedades crónico degenerativas.* Morales-González, JA., Fernández-Sánchez, AM., Bautista-Ávila, M., Vargas-Mendoza, N., Madrigal-Santillán, EO (ed). Ed. UAEH, Pachuca, Hidalgo, México, pp. 93-118.

[34] Nakamura, T., Nawa, K., Ichihara, A. (1984). Partial purification and characterization of hepatocyte growth factor from serum of hepatectomized rats. *Biochem Biophys Res Commun.* Vol. 122, pp. 1450-1459.

[35] Orr, WC., Sohal, RJ. (1994). Extension of life-span by overexpression of superoxide dismutase and catalase in drosophila melanogester. *Science.* Vol. 263, pp. 1128-1130.

[36] Parra-Vizuet, J., Camacho-Luis, A., Madrigal-Santillán, E., Bautista, M., Esquivel-Soto, J., Esquivel-Chirino, C., García-Luna, M., Mendoza-Pérez, JA., Chanona-Pérez, J., & Morales-González, JA. (2009). Hepatoprotective effects of glycine and vitamin E during the early phase of liver regeneration in the rat. *African Journal of Pharmacy and Pharmacology.* Vol.3, No. 8, pp. 384-390 (August, 2009).

[37] Piña-Garza, E., Gutiérrez-Salinas, J., Morales-González, JA., & Zentella de Piña, M. (2003). ¿Es tóxico el alcohol?" In: *Temas Bioquímicos de vanguardia.* Riveros Rosas, H.,

Flores-Herrera, O., Sosa-Peinado, A., Vázquez-Contreras, E. (ed). Ed. Facultad de Medicina UNAM, pp. 121-146.

[38] Ramírez-Farías, C., Madrigal-Santillán, E., Gutiérrez-Salinas, J., Rodríguez-Sánchez, N., Martínez-Cruz, M., Valle-Jones, I., Gramlich-Martínez, I., Hernández-Ceruelos, A., & Morales-González JA. (2009). Protective effect of some vitamins against the toxic action of ethanol on liver regeration induced by partial hepatectomy in rats. *World Journal of Gastroenterology* Vol. 14, pp. 899-907.

[39] Riehle, KJ., Dan, YY., Campbell, JS., & Fausto, N. (2011). New concepts in liver regeneration. *Journal of Gastroenterology and Hepatology.* Vol. 26, Suppl. 1, pp. 203–212.

[40] Rybczynska, M. (1994). Biochemical aspects of free radical mediated tissue injury. *Postepy Hig Med Dows.* Vol. 48, pp. 419-441.

[41] Sandoval, M., Lazarte, K., & Arnao, I. (2008). Hepatoprotección antioxidante de la cáscara y semilla de Vitis vinífera L. (uva) (2008). En: *Anales de la Facultad de Medicina* (citado el 3 de septiembre de 2011). Disponible en: http://www.scielo.org.pe/scielo.php?pid=S1025-55832008000400006&script=sci_arttext

[42] Schandalik, R., Gatti, G., & Perucca E. (1992). Pharmacokinetics of silybin in bile following administration of silipide and silymarin in cholecystectomy patients. *Arzneimittel-Forschung.* Vol. 42, pp. 964-968.

[43] Sohal, RS., Sohal, BH., & Orr, WC. (1995). Mitochondrial superoxide and hydrogen peroxide generation, protein oxidative damag, and longevity in different species of flies. *Free Radic Biol Med.* Vol. 19, pp. 499-504.

[44] Téllez, J., & Cote, M. (2006). Alcohol etílico: un tóxico de alto riesgo para la salud humana socialmente aceptado. *Revista Facultad de Medicina de la Universidad Nacional de Colombia.* Vol 54, pp. 32-47.

[45] Tzu-Chen, Y., Kwam –Liang, K., & Hish-Chen, L. (1994). Age dependent increase of mitochondrial DNA deletions together with lipid peroxide and superoxide dismutase in human liver mitocondria. *Free Radic Biol Med.* Vol. 16, pp. 207-214.

[46] Venereo Gutiérrez, JR. (2002) Daño oxidativo, radicales libres y antioxidantes. En: Revista Cubana de Medicina Militar, Febrero 2002, Disponible en: http://bvs.sld.cu/revistas/mil/vol31_2_02/MIL09202.pdf

[47] Yoshida, Y., Komatsu, M., Ozeki, A., Nango, R., & Tsukamoto, I. (1997). Ethanol represses thymidylate synthase and thymidine kinase at mRNA level in regenerating rat liver after partial hepatectomy. *Biochim Biophys Acta.* Vol. 1336, pp. 180-186.

Permissions

The contributors of this book come from diverse backgrounds, making this book a truly international effort. This book will bring forth new frontiers with its revolutionizing research information and detailed analysis of the nascent developments around the world.

We would like to thank Prof. Dr. José Antonio Morales-González, for lending his expertise to make the book truly unique. He has played a crucial role in the development of this book. Without his invaluable contribution this book wouldn't have been possible. He has made vital efforts to compile up to date information on the varied aspects of this subject to make this book a valuable addition to the collection of many professionals and students.

This book was conceptualized with the vision of imparting up-to-date information and advanced data in this field. To ensure the same, a matchless editorial board was set up. Every individual on the board went through rigorous rounds of assessment to prove their worth. After which they invested a large part of their time researching and compiling the most relevant data for our readers. Conferences and sessions were held from time to time between the editorial board and the contributing authors to present the data in the most comprehensible form. The editorial team has worked tirelessly to provide valuable and valid information to help people across the globe.

Every chapter published in this book has been scrutinized by our experts. Their significance has been extensively debated. The topics covered herein carry significant findings which will fuel the growth of the discipline. They may even be implemented as practical applications or may be referred to as a beginning point for another development. Chapters in this book were first published by InTech; hereby published with permission under the Creative Commons Attribution License or equivalent.

The editorial board has been involved in producing this book since its inception. They have spent rigorous hours researching and exploring the diverse topics which have resulted in the successful publishing of this book. They have passed on their knowledge of decades through this book. To expedite this challenging task, the publisher supported the team at every step. A small team of assistant editors was also appointed to further simplify the editing procedure and attain best results for the readers.

Our editorial team has been hand-picked from every corner of the world. Their multi-ethnicity adds dynamic inputs to the discussions which result in innovative outcomes. These outcomes are then further discussed with the researchers and contributors who give their valuable feedback and opinion regarding the same. The feedback is then collaborated with the researches and they are edited in a comprehensive manner to aid the understanding of the subject.

Apart from the editorial board, the designing team has also invested a significant amount of their time in understanding the subject and creating the most relevant covers. They scrutinized every image to scout for the most suitable representation of the subject and create an appropriate cover for the book.

The publishing team has been involved in this book since its early stages. They were actively engaged in every process, be it collecting the data, connecting with the contributors or procuring relevant information. The team has been an ardent support to the editorial, designing and production team. Their endless efforts to recruit the best for this project, has resulted in the accomplishment of this book. They are a veteran in the field of academics and their pool of knowledge is as vast as their experience in printing. Their expertise and guidance has proved useful at every step. Their uncompromising quality standards have made this book an exceptional effort. Their encouragement from time to time has been an inspiration for everyone.

The publisher and the editorial board hope that this book will prove to be a valuable piece of knowledge for researchers, students, practitioners and scholars across the globe.

List of Contributors

I. Milisav
University of Ljubljana, Laboratory of oxidative stress research, Faculty of Health Sciences, Ljubljana, Slovenia
University of Ljubljana, Faculty of Medicine, Institute of Pathophysiology, Ljubljana, Slovenia

B. Poljsak
University of Ljubljana, Laboratory of oxidative stress research, Faculty of Health Sciences, Ljubljana, Slovenia

Manuel Soriano García
Chemistry of Biomacromolecules Department, Chemistry Institute, National Autonomous University of México, University City, Mexico

Carmen Valadez-Vega, Luis Delgado-Olivares, José A. Morales González, Ernesto Alanís García, Esther Ramírez Moreno, Manuel Sánchez Gutiérrez, Zuñiga Pérez Clara and Zuli Calderón Ramos
Institute of Health Sciences, Autonomous University of Hidalgo State, Ex-Hacienda de la Concepción, Tilcuautla, Hgo, Mexico, C.P.42080., Mexico

José Roberto Villagomez Ibarra
Institute of Basic Sciences, Autonomous University of Hidalgo State, Km 4.5 Carretera Pachuca-Tulancingo, Ciudad del Conocimiento, Mineral de la Reforma Hidalgo, C.P. 42076, Mexico

María Teresa Sumaya Martínez
Secretary of Research and Graduate Studies, Autonomous University of Nayarit, Ciudad de la Cultura "Amado Nervo", Boulevard Tepic-Xalisco S/N. Tepic, Nayarit, Mexico

José Luis Silencio Barrita
Association of Chemists, National Institute for Medical Sciences and Nutrition "Salvador Zubiran", Mexico

María del Socorro Santiago Sánchez
Department of Nutrition and Dietetics, General Hospital No, 30 "Iztacalco", Mexico

Maria de Lourdes Reis Giada
Department of Basic and Experimental Nutrition, Institute of Nutrition, Health Sciences Center, Federal University of Rio de Janeiro, Brazil

Eva María Molina Trinidad, Sandra Luz de Ita Gutiérrez, Ana María Téllez López and Marisela López Orozco
Universidad Autónoma del Estado de Hidalgo UAEH, Instituto de Ciencias de la Salud ICSa, Área Académica de Farmacia, ExHacienda la Concepción, Tilcuautla, Hidalgo, México

José A. Morales-González, Evila Gayosso-Islas, Cecilia Sánchez-Moreno, Carmen Valadez-Vega and Eduardo Madrigal-Santillán
Instituto de Ciencias de la Salud, UAEH, México

Ángel Morales-González
Escuela Superior de Computo, IPN, México

Jaime Esquivel-Soto, Cesar Esquivel-Chirino and Manuel García-Luna y González-Rubio
Facultad de Odontología, UNAM, México

Printed in the USA
CPSIA information can be obtained
at www.ICGtesting.com
JSHW011403221024
72173JS00003B/402

9 781632 395047